WOODLICE

INVERTEBRATE TYPES

WOODLICE

By
STEPHEN SUTTON, M.A., D.Phil.
Lecturer in Zoology
University of Leeds

Key written in collaboration with
PAUL HARDING
DAVID BURN, B.Sc.

PERGAMON PRESS

OXFORD · NEW YORK · TORONTO · SYDNEY · PARIS · FRANKFURT

U.K.	Pergamon Press Ltd., Headington Hill Hall, Oxford OX3 0BW, England
U.S.A.	Pergamon Press Inc., Maxwell House, Fairview Park, Elmsford, New York 10523, U.S.A.
CANADA	Pergamon of Canada, Suite 104, 150 Consumers Road, Willowdale, Ontario M2J 1P9, Canada
AUSTRALIA	Pergamon Press (Aust.) Pty. Ltd., P.O. Box 544, Potts Point, N.S.W. 2011, Australia
FRANCE	Pergamon Press SARL, 24 rue des Ecoles, 75240 Paris, Cedex 05, France
FEDERAL REPUBLIC OF GERMANY	Pergamon Press GmbH, 6242 Kronberg-Taunus, Hammerweg 6, Federal Republic of Germany

Originally published in 1972 by Ginn & Company

This edition 1980

Sutton, Stephen
Woodlice.
1. Oniscoidea
I. Title
595'.372   QL444.M34   80-41268
ISBN 0 08 025942 1

Printed in Great Britain by A. Wheaton & Co. Ltd, Exeter

PREFACE

I have written this introduction to the biology of woodlice because I feel that the increasing use of these animals in biology teaching has created a demand for more background information than has hitherto been available to the people who want it. The use of these animals in teaching has also underlined the need for a straightforward and well-illustrated key which can be used to identify immature as well as adult stages. One further reason for putting pen to paper is that I find woodlice to be very interesting animals and hope to convey this interest to other people. The book is mainly intended for sixth-form staff and students, but I hope that anyone with a bent for biology and an enquiring mind will be able to follow the text and use the key. Also, to satisfy the needs of those who wish to follow up the topics mentioned, I have given a selected bibliography, the use of which will open up large areas of information which the need for brevity has forced me to omit.

Woodlice are eminently accessible and obliging animals, very useful for demonstrating a range of biological principles and so very suitable for experimental work. For these reasons I have included at the end of the book a number of investigations and suggestions for further work to supplement those already at large in the educational literature. Among other things, the successful completion of these projects depends upon accurate identification of species, and so no pains have been spared by myself and my collaborators to produce a key which can be used successfully by beginners. The task is not easy so far as woodlice are concerned but, by concentrating on the commoner species and giving plenty of figures, I believe that the difficulties have been minimised. All along I have been fortunate in having a zoologically qualified artist of considerable talent. The colour plates, I think, bear this out. They were all drawn from life and are, I believe, the first colour illustrations of woodlice (apart from some studies of doubtful quality by Jan van Kessel and other seventeenth-century painters of Dutch Still Lifes). I would like to thank Hilary Burn for all her precise and satisfying artwork.

I should also like to thank the many people who have helped to produce this book. For the use of unpublished information I am grateful to Dr D. T. Anderson, Dr Robin Bedding, Dr Hans-Eckhard Gruner, Mr John Metcalfe, Dr Oscar Paris, and Dr Christopher Rees. For the gift and loan of material my thanks are due to Dr Gruner, Mr Paul Harding, Dr Roger Lincoln, and Mr Peter Skidmore, and for the use of Stereoscan facilities in the Department of Textile Industries of Leeds University, to Dr J. Sikorski. Mr Arthur Holliday prepared figure 8.

I am particularly grateful to Mr David Burn for reading the whole of the main text and for his detailed criticism. My thanks are also due to Dr Lincoln for proof reading the main text, and to Mr Peter Calow, Mr Paul Harding, Dr David Holdich, and Dr Bryan Shorrocks for criticising various parts of the manuscript. None of these people is in any way responsible for views expressed or errors committed. I must thank Dr Charles Fletcher for devising the experimental procedure in Investigation 4, Dr John Gray for providing me with the basis for Investigation 2, and Dr T. Kerr for histological advice. For criticism of the Investigations I owe thanks to many people, but particularly to Dr Sheila Gosden and Mr John Metcalfe. My wife has been a great help at all times and, being the only person who can read my writing, typed the manuscript.

S. L. Sutton

CONTENTS

INTRODUCTION

Woodlice are one of the very few land-living groups of the class Crust-acea. They are more closely related to crabs, lobsters, and water fleas than they are to other terrestrial arthropods such as insects, spiders, and centipedes. Woodlice belong to the order Isopoda—a large group of animals that otherwise live in the sea, or in fresh water. Most of these aquatic forms have given up the swimming habit of the more primitive Crustacea in favour of a bottom-living or parasitic existence; their seven pairs of thoracic limbs, therefore, are designed for walking rather than for swimming and are rather uniform in structure. It was because of this uniformity that these animals were called Isopoda (from the Greek *isos* meaning 'equal' and *podes* meaning 'feet').

Woodlice themselves are thought to have evolved from ancestors that lived on the sea-bed, by way of intertidal forms which developed the ability to survive out of the water for longer and longer periods. The most primitive kinds of woodlice, *Ligia* for example, have evolved very little from this intertidal condition. They inhabit the upper reaches of the sea-shore rather than fully terrestrial habitats, and can stand immersion in salt water indefinitely. Evidence from fossils throws little light on the problem of when isopods emerged onto land, the fossil record being very incomplete. Marine isopods are known to be very ancient, but the earliest woodlice fossils appear only in Eocene deposits, laid down 50 million years ago. We have good reason to believe that woodlice evolved very much earlier than this because all the main families have a world-wide distribution and must have evolved before the continents drifted apart in Mesozoic times about 160 million years ago.

Terrestrial isopods, or woodlice, are all included in the sub-order Oniscoidea of the Isopoda, on the assumption that they have a common terrestrial ancestor. However, current opinion is that they represent three separate invasions from the sea, one being largely a failure (the Tylidae, not found in Britain), one being quite successful—giving rise to the large family Trichoniscidae—and a third (stemming from a form similar to *Ligia*) giving rise to all the slaters and pillbugs represented by *Oniscus*, *Porcellio*, and *Armadillidium*, with their many relations (Vandel, 1965). Because there are no really old fossils to illustrate the evolutionary pathways, these opinions are based on a comparative study of the structures of living marine and terrestrial forms. Such a comparative study reveals the curious fact that the more primitive woodlice closely resemble existing marine isopods in body form, even though they live in such different environments, and there can be little doubt that the ancestors of modern terrestrial forms were in a sense 'pre-adapted' for a successful invasion of the land. This is to say that they had, in the process of becoming adapted to life on the sea-floor, evolved the very features essential for the transition from sea to land.

There are several major problems affecting body structure which have to be overcome when making this transition. For instance, animal tissues are buoyed up in water and the body needs little support; on land, however, a strong skeleton is needed, as well as sturdy legs to hold the body off the ground if rapid movement is required. Filter-feeding, so widespread among aquatic animals is, of course, impossible on land; furthermore water is not available to transfer sperm, so internal fertilization is necessary. At later stages in the life history there is the problem of protection from water-loss faced by all land animals.

As far as the marine ancestors of woodlice are concerned, clearly the walking limbs, which fitted them for a bottom-living (or benthic) existence, with some strengthening, gave the mobility required on land. At the same time, the dorso-ventral flattening of the body (from top to bottom) typical of isopods lowered the centre of gravity and improved stability. Their close relatives, the amphipods (or sand-hoppers) are laterally compressed (from side to side) and have a high centre of gravity. They have been much less successful on land, perhaps because they are so unstable. In taking up a benthic existence, marine isopods abandoned filter-feeding and developed biting and chewing mouth-parts for handling bulky food-items found on the sea-bed. Again, this adaptation proved well-suited for life on land. A further important feature of the marine forms is that they practise internal fertilization, the abdominal limbs of the male being modified to ensure transfer of sperm. Thus, there was no compulsion to return to water at breeding times, as must other land animals handicapped by external fertilization.

Independence from water during breeding was also made possible by the *brood pouch* in which the young could grow while protected from all the dangers of water-loss. The brood pouch probably developed in marine forms as a means of reducing the mortality-rate of the defenceless young, but on land it also has the advantage of protecting them from desiccation. After fertilization in the body, the eggs are laid in a cavity formed beneath the thorax by overlapping outgrowths of cuticle. In this cavity, filled with water, the embryos grow and, after hatching from the egg membranes, continue their development. On release, they are more mature and better able to fend for themselves than if they had been set free just after hatching. The mortality-rate in the brood pouch is very low and the arrangement is clearly of great value in making the transition from sea to land.

However, it would be quite wrong to give the impression that this and other structural pre-adaptations were sufficient to guarantee success on land. For one thing, except for the young animal in the brood pouch, the modifications offer no solution to the most severe problem of life on land—desiccation. Water is necessary to maintain the vital activities of the body and is being lost continually, particularly through transpiration from the body surface. One of the key requirements for minimizing this water-loss is to have a waterproof cuticle, so that transpiration can be limited to the main respiratory surfaces. Further saving can be achieved if diffusion between the areas of oxygen-uptake and the outside world can be kept to the minimum required for metabolic processes. Insects, acknowledged masters of the terrestrial environment, score on both points. Their cuticle has a waxy, waterproof layer and the respiratory system consists of *tracheae* which can be isolated from the air outside by means of spiracular valves. Woodlice, on the other hand, respire by means of modified gills and (in the highly evolved forms) by rudimentary tracheal systems, neither of which has controlled openings. Furthermore, only in the most terrestrial species does the cuticle present any great barrier to the evaporation of water, and even in these species the cuticle is much less waterproof than in the typical insect. Insects have also found more effective means of extracting water from faeces than woodlice have, although the latter have been shown recently to be more efficient than originally thought.

Because of their limited ability to conserve water, woodlice would be fitted only for a very limited exploitation of land habitats were it not for the possession of simple but effective patterns of behaviour which prevent them from staying too long in areas of low humidity. In those woodlice which feed in the open, behavioural reactions confine their activity to the hours of darkness (when humidity is generally higher) and guide them to moist refuges to spend the daylight hours in safety. Thus, in spite of their limitations, woodlice have been able to colonize a wide variety of terrestrial habitats and occur in surprisingly dry situations. They also show a wide range of adaptation, from primitive types such as *Ligia* and *Ligidium* which are poorly adapted to terrestrial life because they lose water rapidly, through the Oniscidae and Porcellionidae, to the pillbugs, *Armadillidium* for example, which are quite well-adapted and flourish in dry, even desert, conditions.

It is often said of woodlice that they are unsuccessful as land animals, being restricted by shortcomings of structure and physiology to a narrow range of dank and dismal habitats in which they live a precarious existence under the constant threat of desiccation. In fact, as I hope to show in the chapters that follow, most of the recent evidence suggests that woodlice are by no means unsuccessful as land animals. There are certain important types of habitat to which they are extremely well-adapted and in which they play a significant part as members of the biological community.

I   BODY STRUCTURE

Like all Arthropoda, woodlice are segmented animals with a rigid exo-skeleton and jointed limbs. There are three groups of segments, the first being the head, the second the thorax or *pereion*, and the third

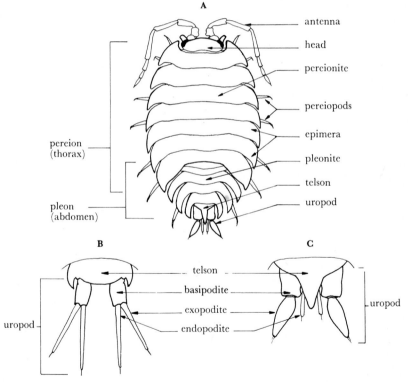

Fig. 1 **A** Dorsal view of *Oniscus asellus* showing major external features; **B** dorsal view of telson and uropods of *Ligia oceanica* and **C** *Porcellio scaber*.

the abdomen or *pleon* (fig. 1). It is important to realize when using the terms thorax and abdomen that these do not correspond either in function or in numbers of segments to the thorax and abdomen of insects, and this is why the alternative terms are often used. In the early embryo, the segments closely resemble each other but, thereafter, they develop along different paths, the limbs becoming variously modified as antennae, mouthparts, walking legs, or gills, depending on the part of the body in which they occur.

In ancestral isopods the head almost certainly contained 5 segments, the thorax 8 and the abdomen 6 but, in later forms, the most anterior thoracic segment fused with the head, leaving the thorax with only 7 segments. In the head, the segments have become so closely associated that they can no longer be distinguished individually, whereas the thoracic segments are all separate, and there is no carapace as in crabs and lobsters.

External structure

Starting with the head, the large dark eyes immediately attract attention. In most families (for example *Ligia*, plate 2) they are compound and are composed of many *ommatidia* (units) which build up a mosaic image of the surroundings. Not all woodlice have compound eyes—the family Trichoniscidae have groups of up to three *ocelli* (simple eyes; fig. 23).

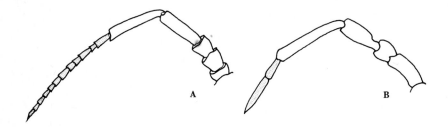

Fig. 2 Dorsal view of left antenna of **A** *Ligia oceanica* and **B** *Armadillidium vulgare*. Segments of the flagellum are shaded; the remaining segments constitute the peduncle.

There are two pairs of antennae. The first pair are vestigial and hard to see and nestle between the second pair, which are large, with many joints. The second antenna is composed of a 5-section peduncle with a multi-articulate flagellum at its tip (fig. 2). The flagellar sections show a progressive reduction in number from *I igia*, which has a great many, through oniscids, which have 3, to the pillbugs, which have only 2. The number of sections, and their lengths, have been much used in classification, but it has to be remembered that these relative lengths change as the animal grows and identification keys with this kind of character cannot be used for juvenile animals. Anybody who has ever watched a woodlouse walking along prodding the ground ahead with its antennae will feel sure that they are sensory organs, but it is very difficult to provide physiological evidence that this is so.

penicils

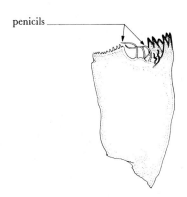

Fig. 3 *Oritoniscus flavus*; mandible of the left side showing the 3 penicils.

    On the under-side of the head lie the mouthparts, which are the modified limbs of the head segments (the antennae also form part of this series). The mouth cavity is enclosed by a *labrum* (upper lip) which is hinged to the head. The most anterior of the mouthparts proper are the mandibles, which have heavily sclerotized teeth, and an articulated spur called the *lacinia mobilis*. The 2 mandibles are dissimilar in shape and have a number of *penicils* (stiff hairs; fig. 3) just below the cutting edge. These vary in number between species and have been used in taxonomic keys, although they are tedious to dissect out and mixing up the left and right mandibles leads to total confusion. Two pairs of maxillae and a pair of maxillipeds come next, and the latter form the 'floor' of the mouth cavity, as can be seen in Figure 4. In feeding, the maxillae and maxillipeds hold the food particles and abrade them so that the mandibles can find a purchase and bite chunks off with the heavily sclerotized cutting edges.

    The maxillipeds are the appendages of the original first thoracic segment now incorporated in the head. They are more or less fused in the mid-line and bear a marked resemblance to the labium of the cock-roach. In fact, the mouthparts of both isopods and amphipods bear a striking resemblance to the basic insect biting type—a good example of convergent evolution towards biting and chewing mouthparts.

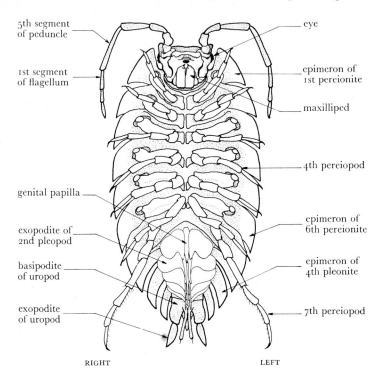

5th segment of peduncle

1st segment of flagellum

genital papilla

exopodite of 2nd pleopod

basipodite of uropod

exopodite of uropod

eye

epimeron of 1st perecionite

maxilliped

4th perciopod

epimeron of 6th pereionite

epimeron of 4th pleonite

7th perciopod

RIGHT   LEFT

Fig. 4 *Oniscus asellus*; ventral view showing major external features. *Note:* the terms right and left are widely used throughout the illustrations; they always refer to the *animal's* right and left sides.

Each segment of the pereion consists of a *tergite* (dorsal plate) and a *sternite* (ventral plate). The lateral edges of the pereion in species such as *Oniscus asellus* project out a long way, forming *epimera*. The *pereiopods* (legs; fig. 4) have a constant number of segments based on the standard crustacean pattern. (See Russell-Hunter, 1969, for their names and for other details of crustacean structure.) The tergites of the pereion and pleon are called pereionites and pleonites respectively.

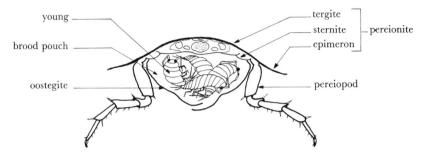

Fig. 5 *Ligia oceanica*; pregnant female with brood pouch, seen in cross-section.

Woodlice have a characteristic stance (fig. 5) which gives them tremendous stability. As with the mouthparts, the development of this form of support by marine forms was an essential preliminary to colonization of land habitats. In the pillbugs, the legs are adapted for burrowing and are stubby with a limited travel. Limbs of this type develop great power and push the animal through the soil. Other woodlice, notably *Philoscia* and *Ligidium*, have developed very long legs and slim bodies for fast running. The trends are the same as those seen in a much more highly developed form in millipedes (burrowers) and centipedes (runners).

The pleon (fig. 1) is always much shorter than the pereion and ends in a pointed telson. The limbs of the abdomen or pleon are the only ones to retain the *biramous* (double-branched) nature so characteristic of the Crustacea. The outer branch of each limb is known as the *exopodite*, and the inner branch the *endopodite*. These abdominal limbs, or *pleopods,* are modified as gills; the exopodites form protective covers which can be raised or lowered, the endopodites form the gills themselves. The first 2 pairs of pleopods are further modified in the male to form external genitalia used in sperm transfer and are very important in identifying species and in sexing animals. The exopodite of the first male pleopod is little modified, but the endopodite has a backwards-pointing projection which is often distinctively shaped or decorated (fig. 6). It forms a pair with the projection of the endopodite on the other side and is also closely associated with the genital papilla projecting from the mid-line of the body. The second pair of male exopodites, like the first, is little modified, but the endopodites are even more elongated to form a second pair of stylets. These two pairs of projections together form an impressive array of sexual equipment. There are no external genitalia in the female.

17

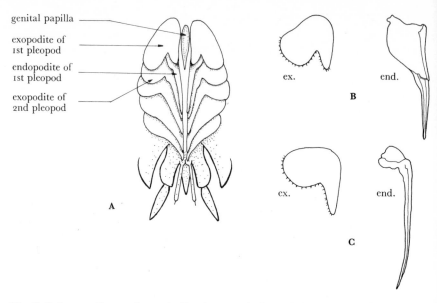

genital papilla
exopodite of
1st pleopod
endopodite of
1st pleopod
exopodite of
2nd pleopod

ex.   end.

B

ex.   end.

C

A

Fig. 6 *Oniscus asellus*; male genitalia, **A** ventral view of the pleon showing the genitalia in position; **B** exopodite and endopodite of the 1st pleopod of the right side; **C** exopodite and endopodite of the 2nd pleopod of the right side.

Another important way in which the pleopods have been modified is in the development of a mass of minute air-channels in the expodites connecting to the outside through a small pore. These air-channels are found only in the more terrestrial woodlice, where they seem to cut down the water-loss associated with oxygen uptake. In some ways they resemble insect tracheae, although they differ greatly in not having a spiracular closing mechanism or direct contact with the tissues where the oxygen is used. Instead, the oxygen has to be carried in the blood.

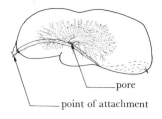

pore
point of attachment

Fig. 7 *Porcellio scaber*; exopodite of the 1st pleopod of female to show pseudotracheal field.

18

Because of these differences, and the fact that they evolved quite independently, they are known as *pseudotracheae* (fig. 7). The presence of pseudotracheae in live animals is easily checked by turning them over and looking for white patches under the pleon. These should not be confused with the white patches visible under the pereion when a moult is imminent. Note that pseudotracheae are not easily visible in specimens which have been preserved in alcohol. The final pair of appendages of the pleon—the uropods—are quite different from those in front.

In *Ligia* (the sea slater), a very primitive woodlouse, the endopodites and exopodites of the uropods are tubular. In higher forms, however, there is a tendency for the exopodites to become flattened and blunt, although the endopodites retain their tubular nature (fig. 1). The exopodites have pores which can exude a secretion repellent to predators. They may also have a sensory function. The body ends in the *telson*, a flattened plate consisting of several segments fused together, often drawn out into a point.

Structure of the exoskeleton
The exoskeleton of the higher Crustacea has been investigated carefully but it is uncertain how closely the general description fits woodlice, on which there have been few direct studies. However, the following features are almost certainly present. The innermost layer is the epidermis, a layer of living cells responsible for secreting the lifeless cuticle overlying it. Above the epidermis lies a thick endocuticle, and above that a very thin surface layer or epicuticle. The main substances found in the endocuticle are *chitin* (a polysaccharide), a protein called *arthropodin*, and calcium carbonate.

Chitin is a very tough but soft and flexible material; skeletal plates are made rigid by the deposition of calcium carbonate and the tanning of the protein to form sclerotin. The intersegmental membranes remain pliable because there is no sclerotinization or calcium carbonate deposition in these areas. Primitive woodlice lack any form of waxy, waterproof layer in the cuticle such as that which has enabled flying insects to become so successful in drier land habitats, but in the more advanced forms an endocuticular wax layer may well exist, while in the desert-living species *Venezillo arizonicus*, there is some evidence for an epicuticular layer resembling that found in insects (see discussion in Edney, 1968). These discoveries suggest that the isopod cuticle is a good deal more versatile in its evolution than has hitherto been realized.

The surface of the cuticle is never smooth, but shows a bewildering

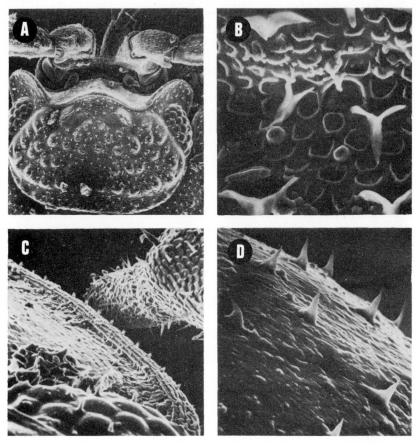

Fig. 8 Details of surface structure as seen with a scanning electron microscope. **A** Head of *Porcellio scaber* (x 17); **B** detail of dorsal surface of pereion of *P. scaber* showing tricorns and other structures (x 580); **C** edge of epimeron of first pereion segment and an antenna of *Armadillidium album*. The facets of the eye can be seen at the bottom of the picture (x 167); **D** dorsal surface of the pereion of *Trichoniscus pygmaeus* (x 1680).

variety of structures especially when studied with a scanning electron microscope. Certain types of structure occur in a wide range of species. A three-pointed tricorn and several other structures are found in *Porcellio* (fig. 8). *Ligia* and the trichoniscid *Androniscus dentiger* have spikes projecting from tubercles, while *Armadillidium vulgare* is unusual in having a pitted surface. The most intriguing structures found so far are those on the rare little *Haplophthalmus mengei* which has a series of dendrites, each surrounded by a polygon of crystals (Sutton, 1969). Treatment with a substance which removes calcium carbonate causes the polygons to disappear. It has no effect on the other cuticular structures, although it does cause a general collapse of the body. The function of these microscopic surface structures in woodlice is quite unknown and, although it is natural to wonder if they are sense organs—perhaps sensitive to touch—they bear no resemblance to sense organs found in other arthropods. Jans and Ross, in 1963, described some features seen in cross-sections of cuticle which they thought might be humidity receptors, but it has not yet been possible to link these structures with anything visible on the surface.

Internal structure

The nervous system is on the standard arthropod plan, with a pair of ganglia above the oesophagus receiving nerve tracts from the eyes and antennae. These ganglia are connected by commissures to the sub-oesophageol ganglia, from which the ventral nerve cord runs the length of the body. The cord is double, with a more or less fused pair of ganglia in each pereion segment and a large fused pleon ganglion.

The digestive system is interesting and is the subject of active study (Schmitz and Schultz, 1969; Holdich and Ratcliffe, 1970). In woodlice, the gut is straight and passes from the oesophagus into the the *proventriculus* (fig. 9). This is a complicated and well sclerotized structure which grinds the food and filters off the juice and small particles, passing them to the four large lobes of the *hepatopancreas* (digestive glands). The latter, with a small ring of cells between the proventriculus and the hindgut, is the only part of the digestive system of endodermal origin and hence is properly called midgut (Goodrich, 1939). All the long stretch of gut between the proventriculus and the rectum is ectodermal in origin and is hindgut, although it is often mistaken for midgut. The hindgut itself consists of a long anterior portion possessing a typhlosole, and a short rectum which opens *via* the anus underneath the telson. Structures in the hindgut which might possibly be involved in water absorption have been found (Smith *et al.* 1969), and some food absorption almost certainly takes place here also.

Little is known about organs concerned with excretion in woodlice, but in related Crustacea, maxillary or antennal glands serve as organs of salt-balance. Clearly, for a land animal the problems of salt-balance are quite different from those facing marine animals, in which these glands have been studied. Thus, one is not sure what to expect in wood-

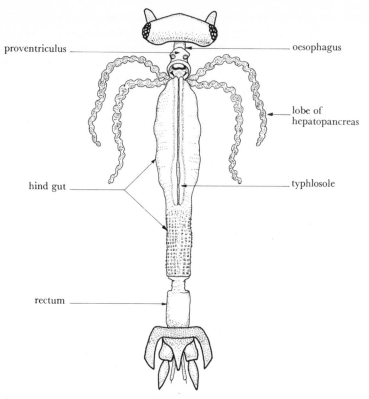

proventriculus — oesophagus

lobe of hepatopancreas

hind gut — typhlosole

rectum

Fig. 9 *Oniscus asellus*; dorsal view of the dissected digestive system.

lice. However, it may be that the hindgut is involved (Smith *et al.* 1969).

The reproductive system is very simple, consisting in the male of a pair of trilobed testes (fig. 10) with a common duct to the genital papilla. In the female, a pair of ovaries opens through the oviducts into the brood pouch (when present) on the underside of the fifth pereion segment. Although some woodlice, *A. vulgare* for example, can store sperm for some months, there does not appear to be a spermotheca. Presumably the sperm is stored in the walls of the oviduct.

The circulatory system is of the standard arthropod type with a dorsal heart and aorta pumping blood forwards and laterally along an elaborate series of arteries into *haemocoeles* (blood spaces) in which the tissues and organs lie. In both terrestrial and aquatic isopods the

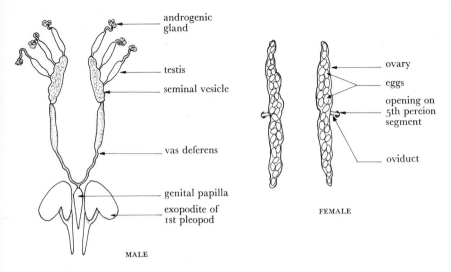

Fig. 10 *Oniscus asellus*; dissections of the male and female reproductive systems.

development of the pleopods as gills has led to a shifting of the heart to the rear of the body, as can be seen clearly in *A. dentiger*, a bright pink species with a transparent cuticle (plate 5). The blood drains back through the blood spaces towards the gill area where it is reoxygenated. It then passes through blood vessels back to the heart.

The internal space of the pereion and pleon is largely taken up by the gut and the hepatopancreas, the testes or ovaries (during the breeding season), the muscles of the legs, and, in some species, the *tegumental glands*. These glands, in an adult *Oniscus*, may occupy one-third of the total body space. Their function has been much disputed, but Gorvett, in 1956, showed that the lobed type secretes fluid which repels predators. There is a pair for each segment (the lateral plate glands) and a group which open through the uropods. The latter glands are well developed in many species, extending well forward into the pereion, whereas the lateral plate glands are always small. Other tegumental glands appear to secrete substances necessary for hardening the cuticle after moulting (Stevenson, 1961).

Embryology and development
In the breeding season, gravid females form a brood pouch, moulting especially for this purpose. The pouch is made up of overlapping

oostegites (leaves) attached to the sternites of segments 2–5 at the base of the legs and projecting inwards to form a 'false floor' to the body (fig. 5). The space is filled with liquid and into it pass the fertilized eggs through the 2 oviducts.

The embryonic development of woodlice has recently been studied in detail by Dr D. T. Anderson of the University of Sydney, to whom I am very grateful for notes on this unpublished work, from which I have compiled the following brief summary:

All the work was done on embryos of *P. scaber* cultured in the laboratory at 21 °C (see Investigations, p. 115, for practical details).

The egg starts as a large globular mass of yolk enclosed within 2 membranes, the inner one tight-fitting, the outer one loose. The reason for having 2 membranes rather than 1 is not known. During the first $2\frac{1}{2}$ days, the egg passes through distinct stages with 2, 4, 8, and 16 nuclei which gradually migrate from within the yolk-mass to collect in a group on the surface of the egg at one pole. This polar mass proliferates and after the 8th day, begins to elongate by adding on cells at the rear. At the same time the first signs of segmentation appear at the front end. By the 17th day, the limb-buds are visible on the pereion and pleon segments, and the hepatopancreas has begun to form. All this time the embryo gradually increases in size until it bursts the outer membrane. The inner membrane remains intact because it is particularly elastic. Meanwhile, the remaining yolk passes into the digestive glands, the head begins to differentiate and, by the 21st day, the first slow movements of the body can be seen.

Up until now the animal has been bent round in the egg with its ventral surface outermost, but it now rotates on its long axis through 180° (like a chicken on a spit) so that its ventral surface is on the inside and the animal can more easily grow into the normal woodlouse shape. Development within the remaining egg membrane goes on until the 26th day, by which time the embryo has darkened (indicating secretion of the cuticle) the eyes have become pigmented, and the yolk supply is exhausted. Increasing movement leads to rupture of the remaining membrane. The young escape into the brood pouch where they remain for several days, during which time the fluid in the pouch gradually disappears.

Hatching from the brood pouch seems to be simply a matter of the young crawling out when they are ready. At first, they have only 6 pereion segments but, at the first moult, they produce a 7th. At the second moult this segment develops a pair of legs to give the full complement of limbs. Thereafter, there are no distinctive morphological differences between moults, which is why intermoult stages in woodlice are called *stadia* as opposed to *instars*, which are stages with character-istic morphological features. Detailed descriptions of the growth stages and development rates of the commoner woodlice will be found in Heeley (1941).

2 PHYSIOLOGY

After years of comparative neglect, the physiology of woodlice is now the subject of much active research. Present interest centres largely on the ways in which woodlice have met the physiological problems that go with life on land. The chief of these is maintaining the correct water-balance so that salts and other substances in the body-fluid become neither too concentrated nor too diluted for life to continue.

Except for *Ligia*, woodlice have little ability to regulate the concentration of the *haemolymph*. Instead, they have to keep the dilution within tolerable limits by moving around until they find environmental conditions which suit them. They have also become osmo-tolerant and are able to withstand quite a wide range of dilutions before death intervenes.

Water uptake

Woodlice have been shown to absorb water (Spencer and Edney, 1954) in the following ways: 1, with the food; 2, by drinking with the mouthparts; 3, by capillary action through the uropods. In addition, there is strong circumstantial evidence that woodlice absorb water through the cuticle at humidities approaching saturation: in fact, they may be unable to avoid it (Den Boer, 1961). Unfortunately, the crucial experiments for establishing water uptake through the cuticle are very difficult to carry out successfully so the exact conditions which cause water overload are not known. What is known is that woodlice *behave* as if they were trying to shed water after a spell in excessively damp or waterlogged habitats, and this can be demonstrated experimentally or seen in the field. Thus, woodlice are often seen climbing up walls and vegetation after heavy rain as if to find drier surroundings. However, much must depend on temperature, because this climbing behaviour is not seen in winter. It is also known that woodlice die if they are kept in wet conditions from which they cannot escape and that total immersion in water proves fatal in the end to all woodlice except *Ligia*, although the more primitive types may survive for several days. Again,

27

much depends on temperature. Death may be caused by water overload, but since survival is better in well oxygenated water, suffocation may also be a factor. In either case it is clear that woodlice, in adapting to terrestrial conditions, have forfeited their aquatic potential to some extent. The realization that woodlice may on occasions suffer from too much water is relatively new—the emphasis in the past being entirely on the problems of water acquisition and conservation. There is scope here for studies of water passage through the cuticle by studying weight gains at very high humidities (95–100 per cent) but, as Spencer and Edney found, there are technical problems in preventing the woodlice drinking through the mouth and anus.

In normal circumstances woodlice can probably obtain all the water they need from their food, but drinking through the mouth and anus are useful where the available substrate is moist and food is not. Anal drinking is accomplished by pressing the uropods close together to form a capillary tube; water is drawn up and absorbed in some way through the anus. The water also passes through an extensive series of capillary channels which run along the underside of the body and moisten the body surface, particularly the pleopodal gill surfaces.

Water loss

We can distinguish 3 main routes by which water leaves the body: 1, by transpiration through the cuticle; 2, with the faeces; 3, through the uropods (by the reverse capillary action mentioned above). The uropod action is probably little used but could be important where surface water is a problem, as when a small woodlouse is hit by a rain-drop. (This must be like having a large tank of water poured over oneself!) Simulating this in the laboratory using a pipette immediately causes a species such as *Trichoniscus* or *Philoscia* to draw its uropods together and then press them to the floor, which, if dry enough will absorb the water. The effect is best seen using coloured water with filter paper as a substrate.

Studies of the water content of faeces (Kuenen, 1959) make it clear that woodlice are able to extract up to 40–50 per cent of the water taken in with their food. He also found that the ability to produce dry faeces is greater in the more terrestrial species, with *Armadillidium vulgare* showing a greater efficiency than *Porcellio scaber* which, in turn, produced drier faeces than *Oniscus asellus*. For the future, it would be valuable to know how the proportion of water in the faeces varies with the desiccation of the animal, and whether water can actually be added to the gut contents from the body fluid. (Kuenen had a few results which could suggest this.)

Transpiration from the body surface and gills accounts for much of the water loss by land isopods. It is determined by the permeability of the cuticle, the temperature of the body, and the drying power of the air. In addition, because small animals have a larger surface area in relation to volume than large ones, size is important also. It is no accident that small woodlice are much more retiring than large ones.

One practical point about transpiration is that the drying power of air is determined by the amount by which the water vapour in the air falls short of saturation, rather than by the relative humidity, which measures only the amount of water as a proportion of saturation value. For example, air with relative humidity of 80 per cent at 10°C will have much less drying power than air of the same relative humidity at 30°C. At 10°C, 1 m³ of air at 80 per cent RH could accommodate only 1·87 g more water vapour before it became saturated; at 30°C, it could absorb 6·01 g (see Edney, 1957, for further discussion of this point). This should be borne in mind when planning experiments on transpiration.

Recently it has been shown (Lindquist, 1968) that the prevalent idea of the woodlouse cuticle as a simple surface acting rather like a piece of porous pot may have to be extensively revised, because he has found a periodicity in water-loss suggesting that there are changes in the permeability of the cuticle. This work raises a large number of questions, particularly about the way in which permeability is controlled and whether the whole cuticle is affected or just the pleopodal region.

There is a very close link between respiration and transpiration, because the moist surface needed for oxygen uptake and carbon dioxide removal makes transpiration inevitable. In primitive woodlice, 80 per cent of water-loss occurs over the body surface and only 20 per cent across the pleopod gills. But, in higher forms, there is a progressive tendency for respiration (and hence transpiration) to be concentrated in the gill region. Pillbugs and other advanced forms lose far less water than the more primitive types and if, as seems reasonable, one assumes that oxygen consumption is just as great, it follows that these advanced forms in respect of water-loss, are more efficient at gas-exchange. This efficiency is attributed to the presence of pseudotracheae, with the implication that because these have a large enclosed surface with one small external opening, they lose less water than an open surface like a gill. As Edney (1968) has pointed out, a small exterior opening of itself will only slow up the rates of diffusion of oxygen and water, not alter them relative to each other. The point of a small external opening is that it can be closed or narrowed to cut down water-loss when oxygen requirements are not maximal, as for instance when the animal is at rest. The efficiency of the insect tracheal system in conserving water rests with the spiracles which close in such a situation. The problem with pseudotracheae is that no one has yet found any closing mechanism, although it is likely that air-flow over the pleopods is substantially reduced when the body is pressed close to the ground in flattened types like *Oniscus*, or when it is rolled up, as in the pillbugs. Unfortunately, this is not an argument for pseudotracheae but simply for ventral respiration sites. Whatever the mechanism, woodlice with pseudotracheae are much more efficient at oxygen uptake in dry air than those without, so clearly pseudotracheae are a valuable adaptation.

Temperature tolerance

One of the problems faced by aquatic animals invading the land is the extremes of temperature with which they have to contend. Obviously, temperature tolerance is at a premium and it is interesting to see how woodlice have coped with the problem. Lethal limits vary according to species, period of exposure, drying power of the air, and the previous temperature history of the animal. Those species inhabiting desert habitats, such as *Venezillo arizonicus* studied by Warburg (1965), can tolerate temperatures up to 42° C for short periods (say 1 hour), whereas species from damp habitats, such as *Oniscus*, cannot stand much in excess of 35° C, with a lower limit close to 0° C. Both upper and lower limits are influenced by the temperature at which the animal has previously been kept. Thus *A. vulgare* cultured at 30° C can withstand temperatures between 41° C and −0·5° C, but when cultured at 10° C, the tolerated range shifts downwards to between 38·5° C and −3·0° C (Edney, 1968). This shift in survival limits according to temperature history is called *acclimation*, and explains why animals cultured at normal laboratory temperatures promptly die if put out of doors in mid-winter, even though introduced to a colony of the same species.

The upper lethal temperature is higher for a short exposure than for a long one, particularly at low humidities. This is to some extent explained by the cooling effect of the evaporating water, which brings the body temperature several degrees below ambient when transpiration is rapid. But most of the difference is simply because of the rapid rate of water-loss at high temperatures. During long exposure at high temperatures, animals die of desiccation. Thus, at 50 per cent relative humidity, *Oniscus* can withstand 36° C for one hour but only 15° C over 24 hours. At 100 per cent relative humidity there is no evaporative cooling and, since transpiration is minimal, the survival at high temperatures over 24 hours is much better. In this case the lethal temperature for long and short exposures differ by only a few degrees (see fig. 11). The water and temperature relations of woodlice are discussed in detail by Cloudsley-Thompson (1971).

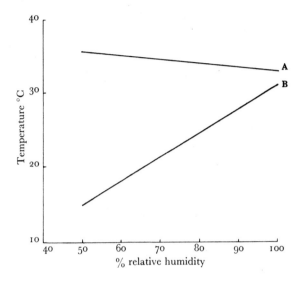

Fig. 11 The relationship between humidity, time of exposure, and upper lethal temperature. The graph shows the highest temperature tolerated by *Oniscus asellus* while exposed to various relative humidities for **A** 1 hour and **B** 24 hours.

Nitrogenous excretion

It is usually stated that the nitrogenous waste-products resulting from protein metabolism (ammonia in particular) are toxic to animal tissues and have to be eliminated from the body at great speed or converted into less toxic substances to keep concentrations down. This statement can be applied to invertebrates only with some caution because it is becoming increasingly clear that quite a number of them have considerable resistance to ammonia toxicity and can tolerate levels of it in their blood which could kill a vertebrate several times over. Notable examples are found among the higher Crustacea, including the isopods.

Animals with a low tolerance to ammonia either excrete it in dilute solution as a copious urine, or detoxify it to form urea or the even less poisonous uric acid. Being less poisonous, these substances can be excreted in more concentrated form than ammonia, with a consequent saving of water. Since for land animals water is always at a premium it is natural to find urea and uric acid dominating the excretory products in these animals. Moreover, cases are known (gastropod molluscs, for example) where the dominant excretory product changes with the dryness of the habitat. Ever since this was established, people have looked to woodlice for a similar series with *Ligia* as an ammonia-producing form, moderately terrestrial species such as *Oniscus* as urea-producing, and the highly terrestrial pillbugs as uric acid-producers.

In fact, no such series has been found. The woodlice tested (*P. scaber* and *O. asellus*) are ammonia-producers and, even more unexpectedly, have no urine; they give off ammonia as a gas. This has come about because of the high tolerance or their tissues to ammonia. Concentrations build up until the rate of diffusion through the permeable cuticle matches ammonia production. At first sight this adaptation might seem to save a great deal of water since none is needed to make urine, but it must be remembered that, hand in hand with diffusion of the gas will go loss of water through transpiration. Thus, whether any water is saved is problematical. Hartenstein (1968) has suggested that the real saving is in energy, because the ammonia is not converted into urea or uric acid. Both these substances are endothermic compounds and use up energy in their formation, so the use of metabolic energy is the price of detoxification. This opens up the intriguing idea that woodlice have a permeable cuticle not because they have failed to come up with anything better in the course of evolution, but because the loss of water is less of a drawback than the loss of energy involved in detoxifying waste products. For the exploitation of terrestrial habitats where water is plentiful, water-loss through the cuticle may be less important than the efficiency with which food is converted into tissue and reproductive products, in which case the less energy used up in excretion processes the better. More information about excretion in pillbugs, which have moved away from moister habitats in most cases, would be very interesting in this context.

The latest development in this story (Wieser *et al.* 1969) is that ammonia is given off in periodic bursts which, at least over most of the year, describe a 24-hour cycle. Two likely explanations for this are variations in the rate of protein metabolism or in the permeability of the cuticle. The latter, of course, has already been suggested by Lindquist as an explanation for variations in water-loss and, as we may expect that the permeability of the cuticle is of the same order for ammonia and water, the two will fluctuate together. The most interesting thing about periodicity of water-loss and gas-loss is that they are inversely proportional to locomotory activity, so that maximum release of the two substances occurs when the animals are at rest in their shelter sites. This could be an adaptation to save water, and would have the incidental effect of releasing ammonia. It would be interesting to know what would happen if an animal became overloaded with water and needed to shed it—one would expect the release of gas and water in this situation to show a positive, rather than an inverse, correlation with locomotory activity. We need to know much more about the whole problem of excretion in isopods, particularly about excretory products in pillbugs.

Digestion

Digestion takes on a special significance in land animals because all the salts and trace elements they need have to be absorbed from the food; in aquatic animals, they can be extracted from the water through the gills. As we shall see, this presents woodlice with special problems.

As it is consumed, the food is mixed with a mucopolysaccharide secreted by tegumental glands in the head (Stevenson and Murphy, 1967). This lubricates the food while the mandibles break it up. It then passes along the oesophagus to the proventriculus, where small particles are pumped into the hepatopancreas, the centre of all digestive and metabolic processes. Here, enzymes are secreted and the broken-down food products absorbed, metabolized, stored, or carried away to other parts of the body. Rapid and rhythmic contractions of the glands ensure that the contents mix and eventually the undigested residue passes into the hind-gut. Whether absorption can take place here, between the stomach and the rectum, is still unsettled but the histology of the region and the presence of a typhlosole (fig. 9) to increase the surface area suggest that it can.

Among the various metabolites stored in the hepatopancreas are large quantities of copper (Wieser, 1966). Curiously, the amount is far in excess of that likely to be needed to keep up the concentration of the blood pigment haemocyanin, and its purpose remains a mystery. What is known is that the animals die if their stocks of copper are exhausted, and that this can happen if they are prevented from eating their own faeces, even when fresh food is plentiful. It appears that sufficient copper to satisfy the animals' needs is present in most types of food but that much of it is bound up in organic complexes which are not broken down in the gut. Instead, they pass out with the faeces

33

where, probably through bacterial activity, they are changed into a form which woodlice can absorb. Thus, a diet of fresh food alone will not allow copper stocks to be maintained but, if faeces can be recycled, the deficit can be made good. It is worth noting that bacterial activity may well make available a lot of other things in the faeces, which may partly explain why woodlice can live for so long in cultures without fresh food.

Respiration

Oxygen uptake from the air has already been described in connection with transpiration, but something can be added on oxygen transport in the body and patterns of energy uptake in relation to metabolic requirements. Until recently it was thought that isopods, along with amphipods, had no respiratory pigment in the blood, but it is now known that, as in most other Crustacea of any size, haemocyanin is present. This is a pigment serving the same function as haemoglobin, but with copper instead of iron in its constitution. It is pale blue when oxygenated and colourless when deoxygenated. It has escaped detection before because the quantities present in isopods are too small to show any tell-tale colour and sensitive techniques are needed to detect it (see Wieser, 1965). How important the pigment is, in view of its low concentration, remains to be seen—much of the oxygen required by the tissues is probably carried in the blood fluid. The respiratory function of both fluid and pigment is more important in those species like the pillbugs, where oxygen uptake is concentrated on the pleopods. These animals clearly place less reliance than primitive types on diffusion to the tissues from the body wall. One also expects that larger animals have a more effective transport system than small ones.

The pattern of oxygen uptake is of some significance because it reflects the amount of energy used in general body metabolism as well as in various forms of activity such as locomotion and feeding. Wieser (see Edney, 1968) found that locomotory activity doubled respiratory rates in *P. scaber* and *A. vulgare*, while the rates also increased sharply in spring with the onset of growth and gonadal development. Phillipson and Watson (1965) also found that breeding activity causes high oxygen consumption and that size is an important factor, juvenile animals having higher respiration rates (relative to size) than all others except breeding adults.

Moulting

Moults were once regarded as short episodes in the life of the woodlouse, interrupting relatively long intermoult periods. It is now realized that

34

the actual shedding of the cuticle is only one of a whole series of events making up the moulting cycle and, except in the depths of winter when there is no growth, or in very old animals which only moult occasionally, an animal will nearly always be in some active stage of the cycle. Events leading up to the process are initiated by hormone secretion from the brain. Among other things, all the copper reserves in the hepato-pancreas are mobilized (though no one knows why), and calcium is withdrawn from the old cuticle and temporarily redeposited in the anterior half of the body in conspicuous white patches along the mid-ventral line. These patches disappear as soon as the rear half of the body has moulted, being used in the strengthening of the new cuticle. Presumably, some is held over to strengthen the front half of the body when this is moulted a few days later. The rest of the cycle is taken up with consolidation and strengthening of the new cuticle. There is some doubt as to whether water has to be taken in (as it has to be in many other arthropods) to expand the soft, new cuticle. Most authors believe that surface water is not necessary, but Standen (1970) working on *Trichoniscus pusillus*, found that if water was not available the animal failed to increase in size at the moult.

Hormone physiology
A great deal is known about a few restricted aspects of hormone function in the higher Crustacea, particularly the control of colour change and of moulting. Very little of this work has been done on woodlice, however, because of their small size and the inaccessibility of their secretory organs. What evidence there is suggests that in most aspects the endo-crinology of woodlice closely resembles that of the better known decapods (shrimps and lobsters); for instance, melanophore control can quite easily be demonstrated in the laboratory using *Ligia* (see Investigations p. 116).

Colour changes in other woodlice are much more difficult to investigate, but have been shown to occur in *Trachelipus* (McWhinnie and Sweeney, 1955).

Although generally having the same arrangements for endocrine control as other Crustacea, isopods (together with amphipods) have a unique process of sex-determination: the development of both the primary male characters (testes) and secondary characters (external genitalia) are controlled by androgenic glands, situated at the free ends of the testes (Charniaux-cotton, 1960). Furthermore, Legrand and others have now shown that implantation of androgenic glands into female woodlice causes male characters to appear. Females develop external genitalia, produce sperm, and even mate with normal females.

35

Androgenic gland activity is triggered by secretion from the nervous system (Legrand *et al.*, 1968).

Sensory physiology

Little work has been done on sensory physiology in spite of the obvious need for such studies. In particular, a search is needed to establish what kind of humidity receptors exist and where they are situated. The difficulties which have prevented progress so far are mainly the small size of the animals and the intricacy of the apparatus needed to record changes in electrical conductivity of receptors. However, Dr C. J. Rees of York University has been doing some preliminary experiments and has established that the antennae of *P. scaber* have sensitive thermoreceptors on the flagella. These could be the long-sought-for humidity receptors if they prove to be sufficiently sensitive to changes in temperature caused by evaporative cooling of the flagella as a result of transpiration.

3 BEHAVIOUR

From what has been written so far it should be quite clear that woodlice have a wide range of structural and physiological adaptations to enable them to survive on land, but only in the damper terrestrial habitats. How do they ensure that they remain in conditions of tolerable humidity, and how do they find their way back to such places after venturing out? The answer lies in their behavioural reactions, which are finely tuned to environmental conditions and beautifully adapted to their physiological needs.

Woodlice respond to unfavourable conditions of the physical environment mostly with simple locomotory responses which have been the subject of much study over the years. In fact, from the amount of attention paid to these responses one might be forgiven for thinking that the behaviour of the group began and ended here although, in reality, there is a whole range of little-known behaviour patterns, such as reaction to predators and to each other. These latter patterns have been neglected because they occur mostly in the depths of the night and tend to be upset or inhibited by the activities of an observer. Aspects of behaviour discussed in this chapter are summarized here:

Locomotory responses to:
1 humidity
2 temperature
3 light
4 solid objects
5 chemical odours
6 wind
7 other stimuli

Other behavioural responses:
1 anal drinking
2 feeding behaviour
3 reproductive behaviour
4 moulting behaviour
5 defensive behaviour

Locomotory responses may be of two kinds: 1, a directional movement orientated to a stimulus (as when a woodlouse runs away from a light); 2, simply a change in level of activity as the intensity of a stimulus changes (as when a woodlouse walks more rapidly as tempera-

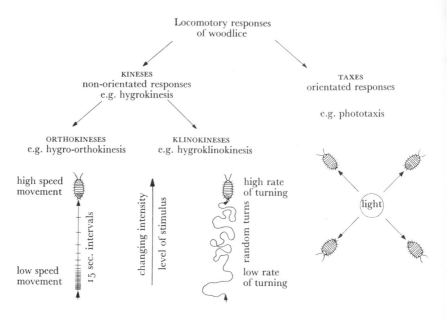

Fig. 12 Types of locomotory responses and their classification.

ture increases). In the terminology of Frankel and Gunn (see Carthy, 1958) a response involving a change in level of activity is called a *kinesis*; an orientated response to a stimulus is called a *taxis*. The terms are summarized in fig. 12. Note that kineses can be classified further into those involving just a change in activity-level (*orthokineses*) and those involving an increase in the number of random turning movements (*klinokineses*). These random turning movements are a feature of woodlouse activity and occur in the absence of orientated (tactic) behaviour. This nomenclature is widely accepted and very useful, although it may be a little difficult to grasp at first.

The response of a woodlouse to a single stimulus such as light, under laboratory conditions, often appears to be clear-cut and stereotyped, but it is very important to realize that, in nature, the strength and type of the response to any one stimulus is influenced by the strength of other stimuli and by the physiological state of the animal. The result is that, in nature, the behaviour of a woodlouse is very varied, responding precisely to external conditions and the needs of the body.

Humidity and temperature differ from other stimuli in that they directly affect the survival of the animal. Lack of humidity or too great a temperature will kill an animal, whereas light or lack of solid objects to rest against cannot (except, perhaps, intense sunlight). However, responses to the latter stimuli serve to lead the animal into areas of favourable humidity and temperature, and play an important part in woodlouse behaviour.

Humidity

The characteristic response to humidity is a kinesis, involving both speed of movement (orthokinesis) and rate of random turning movements (klinokinesis). A desiccated animal, when dropped into a humidity gradient, will show least activity at the damp end of the apparatus and, if conditions are sufficiently moist, will stop moving altogether. In this way, animals tend to congregate in damp places rather than dry ones. However, animals which have been kept in extremely wet conditions react differently. They show an indifference to the humidity gradient or even a tendency to avoid the dampest regions. This last response has been taken as evidence that woodlice become overloaded with water at very high humidities (see also p. 27). Other evidence which can be taken to support this view is the appearance of woodlice such as *Porcellio scaber* and *Oniscus asellus* in the open after heavy rain, and the conclusive demonstration by Den Boer (1961) that woodlice spend more time out in the open on wet nights than on dry ones. Starting with the premise that woodlice absorb water in their shelters during the day and have to come out at night to transpire it, he reasoned that, because on wet nights the air has less drying power, the animals have to stay out longer to transpire their excess water. Observations made by Cloudsley-Thompson (1958) that woodlice are less active on windy nights (because the wind removes the protective 'shell' of moist air around the animal) could also be taken to support this hypothesis. Den Boer carried out very thorough laboratory experiments which supported his view and concluded that activity, at least in *P. scaber* on the study site, was caused by the need to shed water overload and was not motivated by the search for food and mates as is usually supposed.

The orthodox view of the fact that woodlice stay out longer on wet night starts with the premise that woodlice are short of water and that on wet nights their reserves last longer because of the lower transpiration-rates. To resolve this conflict of hypotheses, two vital facts have to be established:

1 What humidities do the animals actually experience in their shelters?

2 Do they actually take up water involuntarily at these humidities? So far, technical difficulties have frustrated attempts to answer these questions, particularly the difficulty of measuring humidity in a confined space at humidities approaching saturation. It must be said, however, that there are plenty of cases known where activity is very obviously not concerned with problems of water balance—for example, where animals are seen to be eating a food not available in shelter sites—and it is doubtful if animals would usually choose as shelters those places which were basically unsuitable from the humidity point of view. At the same time, it seems eminently probable that shelters, on occasions, do become too wet, and that their occupants have to emerge to dry out.

While some, at least, of the activity of the larger woodlice can perhaps be explained as a reaction to water overload, it seems very unlikely that this can be so in the small trichoniscids. Many of these small species seem to live continuously in saturated surroundings and, indeed, soon die if moved from them. *Trichoniscus pygmaeus* loses water so quickly (because of its large surface-area relative to its volume) that it can survive for only a few minutes in a dry Petri dish, and normally never emerges from its home deep in the soil or litter.

As with other responses, the humidity reaction is much modified by the strength of other stimuli and the state of the animals. Thus, at low temperatures (near freezing), humidity responses are slight, while near the upper lethal temperature, the response is reversed: instead of the animal seeking higher humidity it moves towards lower ones and moves out into the open. In this way, it can take advantage of the evaporative cooling effect caused by rapid transpiration to lower the body temperature by a few degrees and thus avoid heat-stroke. This reaction clearly has survival value because it allows animals to move from over-hot shelters to find cooler refuges. Equally clearly it can only be a stop-gap measure because at such high rates of water-loss their reserves will soon be exhausted.

Concerning the effect of light on the humidity response, Cloudsley-Thompson (1952) found that animals kept in darkness have only a weak response to moisture which, he suggests, is valuable in nature because it allows animals to emerge from their shelters after dark even though the humidity is lower outside.

Temperature
Like humidity, heat has a direct effect on the body, and an effective response to it is needed to ensure survival. Above a certain level, increase in temperature has an orthokinetic effect and the speed of

movement rises until heat-stroke and death intervene. However, below a certain level activity again increases. Thus, freezing temperatures in the field cause an increase in activity and result in migration away from the surface layers and into houses and other shelters.

Light
Although light does not affect the physiological state of the animal in the same ways as do humidity and temperature, it plays a very important part in woodlice behaviour. Characteristically, woodlice are negatively phototactic—that is to say, they move directly away from a light source. This normally has great survival value because, in nature, dark places are usually damp places, and so the light response reinforces the humidity response. There are wide differences in the strength of the response in different species such that the less rapid the water-loss the less intensely photonegative the animal appears to be. The pillbugs, which lose water less rapidly than other woodlice, even show a positive phototaxis at high temperature, which explains why they are so often seen about in the open in full sunshine on summer mornings.

Apart from acting as a direct stimulus, light is also the 'clue' by which the intrinsic rhythm of activity in woodlice is kept in step with the cycle of day and night. Cloudsley-Thompson (1952) has shown that woodlice have a peak of activity at night and that this rhythm persists for many days if the creatures are cultured in continuous light. The same rhythm is seen in ammonia release (p. 32). Although it is eventually lost, it may be restored by exposure to the original day/night regime, indicating that day-length is the determining factor in its maintenance. This reliance on day-length for maintaining internal rhythms of the body is general in animals, because few features of the environment are as regular or as easily perceived.

Response to solid objects: thigmokinesis and thigmotaxis
Thigmokinesis is a characteristic response of the cryptozoa (animals living in soil and litter) and has been carefully investigated in woodlice by Friedlander (1963). The response is such that the animal is most active when the contact with the substrate is minimal—that is, when only the feet are on the ground. As soon as other parts of the body touch a surface the animal slows down and may stop if enough of the body makes contact. Thigmokinesis causes woodlice to congregate in crevices between stems of grass or leaves in the litter where they are protected against desiccation and predators. Even other woodlice qualify as solid objects, so that thigmokinesis contributes greatly to the build up of aggregations.

Aggregation is one of the most characteristic forms of woodlouse behaviour and is probably, to some extent, a purely accidental result of individuals acting in the same way to the same stimuli, with thigmokinesis as a prime cause. What biological significance aggregation may have is uncertain and needs investigation. It is known, however, that bunching reduces individual water-loss.

Thigmokinesis tends to supplement humidity and light reactions because crevices and narrow spaces in which it brings woodlice to rest are usually both dark and damp. The strength of the response varies with the desiccation of the animal and is most marked after exposure to dry air (as one might expect) because that is when the animal has most need of damp conditions.

This reaction has sometimes been called thigmotaxis but, as there is no orientation involved, the response cannot be regarded as a taxis. A different response is often seen in choice chambers where the animal, having once made contact with the wall of the chamber, follows it round for some distance. The stimulus appears to be tactile and the response is orientated to the wall. Thus, thigmotaxis appears to be the correct term to use in this case. The nature of the response needs to be examined thoroughly.

Reactions to chemical stimuli

This is a field ripe for investigation because there have been recent developments which throw up interesting possibilities. It has been known for some time that woodlice are sensitive to chemical vapours, choice experiments showing that they are repelled by ammonia and carbon dioxide. A positive response to formic acid has been shown in *Platyarthrus hoffmannseggi*—a small, white, blind species which lives as a commensal in ants' nests, probably feeding on ants' faeces. Originally, the ability to follow a formic acid gradient was hailed as an adaption to allow the species to keep within the confines of the nest until it was realized that some of its hosts do not produce the substance. Clearly, other stimuli are involved—perhaps the same substances as the ants use to recognize each other.

Recently there has been revived interest in the question of whether woodlice produce a scent attractant. Kuenen and Nooteboom (1963) experimented with several species using a choice-chamber and showed that each species tested was attracted to air which had been passed over members of the same species. Acoustic and other stimuli could not be completely excluded, but the fact that species were less responsive when exposed to other species than to their own suggests a species-specific scent. A possible clue to the nature of the attractant is to be found in an unpublished investigation by Dr Robin Bedding on the flies which parasitise woodlice. He discovered that the female parasites were very much attracted to the shelter sites in which wood- lice had been resting, and found good evidence that the attractive factor originated in the uropod gland; he found that pieces of bark impregnated with gland extract were chosen by the parasites as ovi- position sites while untreated pieces were ignored. These glands are usually regarded as part of the defences of woodlice against predation, but it may well be that they have an additional function in 'labelling' good shelter sites, and the parasites have come to use the same scent to locate their victims. Some of the parasites are quite common in places (see Chapter 5) and experiments to explore the importance of uropod gland secretions as a marker substance should not be difficult to devise.

Wind
For an animal in which rate of water-loss is largely determined by transpiration from the body surface, wind is bound to be an important factor because it removes the 'shell' of moist air surrounding the animal. In effect, it drastically lowers the humidity and greatly increases the rate of transpiration. Cloudsley-Thompson (1958) was the first to point this out and observe the consequences of wind. He counted the number of woodlice visible after dark on a stone wall on calm and on windy nights and showed clearly that their numbers fell as wind- speed rose. His conclusion was that wind inhibits activity because of the risk of rapid water-loss and desiccation but, as we have seen, the opposite interpretation—that on windy nights woodlice lose their excess water load so quickly that their activity periods are very short— also fits the known facts quite well.

Other stimuli causing locomotory responses
Prime suspects here are acoustic and gravitational stimuli. Work is needed to assess their importance.

Other behavioural responses
Some of these are very simple—for example, the defensive behaviour of a pillbug which simply rolls up into a ball (albeit a ball of complicated and intriguing design) when molested. But other responses, such as mating behaviour, consist of a sequence of specialized acts building up a complex but characteristic pattern. Locomotory responses nearly always form one component of these complex patterns, as do active movements of the antennae.

Anal drinking and water shedding through the uropods

To take up water from the substrate a woodlouse closes its uropod endopodites together so that they form a capillary tube, then presses them repeatedly onto a water-logged substrate so that water is drawn up by capillarity. Water can be lost in the reverse direction, provided the ground is dry enough to absorb it.

Feeding behaviour

With one known exception, woodlice feed on dead or, at least, immobile materials, so they have no special behaviour to catch and subdue their prey. The exception is *Tylos*, a genus of isopods which live on beaches in the warmer parts of the world. *Tylos latreillei* occurs in the Mediterranean and emerges from the sand at night to chase sand-hoppers, seizing them with its front limbs. Hunting is probably visual because this species has well developed eyes. For other woodlice, finding food is probably a matter of taste and smell. Feeding is never easy to observe since it takes place mostly at night and the animals stop if a light shines on them. In any case, movement of the jaws is very hard to see so that usually one has to rely on the disappearance of food, or the movement of animals onto the food, to indicate feeding.

Mating behaviour

This is very difficult to observe because it usually takes place in total darkness and light disrupts it. I am very grateful to Dr H. E. Gruner of Berlin for my information on this subject.

Woodlice are not known to have any lengthy courtship behaviour as is found in aquatic isopods such as *Asellus*, where the male rides round on the back of the female for some time before mating with her. When a male woodlouse comes across a receptive female (perhaps detected by scent) he stops, tests the air with rapid movements of his antennae, and then brings them to rest on the female. If she does not turn away the male then crawls onto her back (fig. 13A) licking her head with his mouthparts and drumming on her back with his front legs. This goes on for about 5 minutes.

The main phase of mating behaviour begins when the male shifts to a diagonal position on either side of the female (in this case to her left, (fig. 13 B) and bends his body under her so that the *left hand* stylets (endopodites) of his genitalia can reach the *right hand* genital opening on her underside (fig. 13 C). This action seems to be a feat worthy of a contortionist. After 5 minutes or so sperm-transfer is complete and the male crosses over and repeats the performance, this time transferring sperm to the female's *left* genital opening from his *right hand* stylets

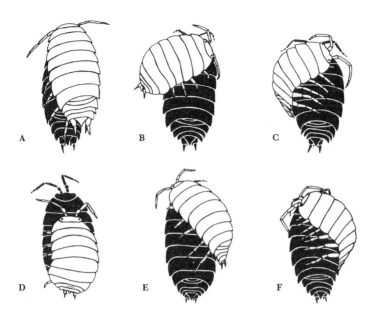

Fig. 13 Mating behaviour in *Porcellio laevis*; the sequence of positions taken up by the male (white) and the female (black) during sperm transfer.

(fig. 13 D, E, and F). Apparently, only the endopodite of the second pleopod is actually inserted, which explains why it is so much longer than the first (p. 17). Mating behaviour is best known in *Porcellio dilatatus* and in *P. laevis*, but is thought to be similar in other woodlice (except the continental Tylidae). However, porcellionids do not have special modifications of the 7th pereiopod in the male, as found in pillbugs and some trichoniscids, and the mating behaviour in these latter forms may be a little different.

Moulting behaviour

A few days before moulting, woodlice stop feeding and become totally inactive. When the rear half of the body is ready to moult the skin splits and the animal drags itself free with its front limbs. A few days later the performance is repeated in reverse when the head end is shed. On both occasions the cast skin is often eaten. In laboratory cultures cannibalism of moulting woodlice by their more mobile neighbours is rife, but one suspects that this is not so in the wild. If it had been, more elaborate moulting behaviour would have developed, like that of

45

Glomeris, a millipede which builds itself a mud coccoon in which to moult. There is, in fact, a continental woodlouse not found in Britain which does construct such a coccoon.

Defensive behaviour

The most characteristic form of defensive behaviour is to keep under cover during the daytime so that predators hunting by sight are avoided. When attacked, woodlice initially respond in one of three basic ways: 1, they run away as fast as possible, like *Philoscia muscorum*; 2, they clamp down on to the surface, like *O. asellus*; 3, they roll up into a ball, like pillbugs.

Each of these basic reactions is linked to a particular body form. Thus, *Philoscia muscorum* is adapted for rapid strategic withdrawal, with a slim body carried on long legs giving the animal a surprising turn of speed. *O. asellus*, on the other hand, has a very flat, oval appearance with a low-slung body. When attacked, the feet grip the surface very tightly, pulling inwards towards the centre line of the body which, at the same time, is pulled down until the edges of the dorsal plates are touching the surface. In this attitude, the animal is remarkably difficult to prise away from its hold unless a long fingernail or sharp probe is brought to bear. Clearly, a predator would find it very difficult to get at the soft parts underneath. Pillbugs are well known for the ability to roll up into a ball (fig. 31) and the body is much modified to make this possible. The animals are arched in shape and grooves have developed in the head into which the antennae fit. Rolling up is usually thought to be a device for limiting water-loss, but you only have to watch the behaviour of a pillbug when put into a cage with a shrew to realize that rolling up is also a defence against predators (just as it is in real armadillos or in the millipede *Glomeris*, which is often mistaken for a woodlouse). When attacked by a shrew, the pillbug, seemingly warned by vibration, snaps shut so that the attacker is unable to find a purchase with its jaws and is reduced to pushing the pillbug around with its nose. Only if the prey is small enough to go into the mouth whole, or if it closes up around a grass stem and cannot shut properly, is the shrew successful.

Apart from these general responses there are others. *Porcellio* species on walls or tree trunks will drop off into the undergrowth when disturbed, while all species will feign death if an attacker persists. Further attack often results in discharge from the uropod glands, as can be seen if *Trichoniscus* or *Philoscia* are put in a Petri dish and harried with a paint brush. After a while they stop moving and droplets of secretion appear on the uropods. This hardens very quickly and makes a very effective gum. These secretions are distasteful to spiders (Gorvett, 1956) and the sticky nature of the substance may add to its deterrent effect. Since starving spiders will eat woodlice we must suppose that the secretions make the prey less appetizing rather than totally repellent.

Spiders are not the only attackers affected; Gorvett and Taylor (1960) showed that the secretions were effective in protecting *Platyarthrus* from the ants with which it lives. They found that if *Platyarthrus* is transferred to a new nest, it is immediately set upon by its new hosts, which try to bite it with their jaws. The woodlouse reacts by clamping down (like *Onsicus*) and, at the same time, turning the tips of its uropods upwards. The ants then bite at these but soon back away rubbing their jaws, apparently trying to remove secretion discharged from the uropods. The glands which provide this secretion are poorly developed in pillbugs, where the ability to roll up offers an alternative means of protection. Much remains to be learnt about these secretions in woodlice—the last work on its chemical composition was that of Gorvett in 1956. Since then, many new analytical techniques have been developed and the identification of the active ingredients might not be at all difficult for someone with access to modern apparatus. In the meantime, a number of experiments could be done to discover the range of predators affected and the deterrent effects of the secretions on them.

Further reading for Chapters 1, 2, and 3:

Edney, E. B. 1954. Woodlice and the land habitat. *Biol. Rev.* **29**, 185–219.
Edney, E. B. 1968. Transition from water to land in isopod crustaceans. *Am. zool.* **8**, 309–326.

4 GENETICS

The genetics of woodlice and other isopods have received very mixed treatment; some aspects have been studied in detail and others virtually ignored.

One feature that has attracted a good deal of attention is the inheritance of body colour and the existence of several distinct colour forms in a single population. This is known as *colour polymorphism* (or *polychromatism*). Many woodlouse species have such colour varieties because of differences in the quantity and nature of the pigments present, and also because of the pattern in which they are laid down. These colour forms either may be connected by a whole series of intermediates, or each variety may be quite distinct from the others. This latter situation is often caused by the presence of several *allelomorphs* (genes operating at the same locus on a chromosome) and studies have been made to see whether this is so in woodlice.

Colour forms

In *Armadillidium vulgare* the normal colour in adults is a dark slaty-grey (almost black in some specimens), but there exists also a distinct rufous form. In a series of breeding experiments Dr H. W. Howard at Cambridge was able to show that there were, in fact, 2 red forms allelomorphic to the normal grey form. They were of identical appearance but different genetic constitution, as shown by the fact that one was dominant to the normal form and the other recessive (Howard, 1962). As might be expected (since recessiveness usually indicates that a character has a deleterious effect), Howard found that the recessive red was rarer than the normal grey form. But what is unexpected is that the dominant red was *also* rarer than the grey form. A possible explanation for this is that the red gene determines other characters besides colour and, while being dominant for colour, it is harmful in some other way, such as, for example, lack of resistance to cold weather.

A comparable situation has been discovered recently by Dr Laura Adamkewicz (1969), working on *Armadillidium nasatum* in Virginia, in the United States but, in this species, the red form recessive to the grey form can occur in up to 36 per cent of the population, and the red form dominant to the grey is very rare. Here we can again suppose that each gene has several effects and that the dominant red is highly deleterious in some other way. But we must also suppose that the recessive red is highly advantageous in some other way and has thus come to be very common in the population.

Red forms occur in many other woodlice but their genetics have not been worked out in any detail. The two main problems in this sort of study are: 1, difficulties of breeding individuals singly and in small groups over several generations; 2, the possibility that sperm remains viable for some months within the female (as certainly happens in *A. vulgare*) so that the only safe course in cross-matings to determine dominance is to rear females in isolation—a long and difficult task.

One species in which breeding difficulties have made for slow progress is *Philoscia muscorum* which, in addition to the normal brown form has a very pretty red form and a yellow-green one (plate 6). The frequencies of these 3 forms differ from one locality to another in Britain, even between sites a few metres apart. Usually the red form occurs at something less than 2 per cent and the yellow-green form is almost absent, but colonies exist in which the red form reaches 10 per cent and others in which the yellow-green form reaches nearly 60 per cent. Large variations in morph frequency over short distances were found by Adamkewicz in *A. nasatum*, and she also found that in each locality the frequency of the red forms decreased during mild winter weather, although the factor causing this could not be identified. The situation is strongly reminiscent of that in the snail *Cepaea nemoralis*, where changes in frequency of the colour varieties in space and time are thought to be caused by climatic effects.

Apart from general colour, woodlice show a wide variety of mottled, blotched, or stripy patterns, the genetics of which (in *Porcellio scaber*) have been studied by Lattin (1954). He revealed a very complex situation. In most species patterning is generally much more marked in young animals and adult females than in adult males, in which pigment deposition is uniform and heavy, obliterating all pattern.

Monogeny
Another aspect of the work done by Howard and Adamkewicz concerns the curious phenomenon known as *monogeny*, in which a female produces a brood which is either entirely male or entirely female, thus ensuring that there is no inbreeding. The extent of monogeny in woodlice has not been fully established, mainly because each brood

has to be cultured for some time before it can be sexed. At least in *A. vulgare* there is also the complication that some females are not monogenetic, but produce bisexual broods, so that the condition may not be detected unless the study is very thorough.

Parthenogenesis

Another genetic eccentricity found in woodlice is *parthenogenesis* (the production of young from unfertilized eggs). It is known to occur in one race of *Trichoniscus pusillus* and has been extensively studied by Vandel (1940). Parthenogenesis is the reverse of monogeny in that outbreeding is not possible because there is no mating, so the gene pool cannot be enriched by the crossing-over of chromosomes.

As a result, beneficial mutations cannot spread through the population which, therefore, cannot adapt as quickly as a normal breeding group to new ecological situations. In the long term, therefore, parthenogenesis is a hazardous enterprise. The great advantage of parthenogenesis is that the male sex, no longer needed to fertilize the females, can be eliminated and the resources of the environment devoted entirely to the production of breeding stock. Having a high rate of reproduction relative to one's competitors means they can be swamped out. Thus, parthenogenesis produces rich rewards in the short term.

Parthenogenesis in *T. pusillus* has another advantage—it has allowed the animals to become triploid. Triploid animals have 3 sets of chromosomes in each cell instead of the normal diploid complement of 2 sets (one set derived from each parent). The diploid chromosome number for *T. pusillus* is 16; the triploid is 24. Triploids, like other polyploids with an uneven number of sets of chromosomes, suffer from failure of the normal processes of fertilization so that, in *T. pusillus*, parthenogenesis (allowing asexual reproduction) must have preceded the condition, or occurred at the same time. The advantage of triploidy (and of polyploidy in general) is that it seems to bring with it increased vigour and ability to resist extreme conditions. Thus, in both animals and plants, there is a notable tendency for polyploid forms to replace related diploid forms as one travels through the temperate zones towards the poles. As one of the very few animals with a polyploid race, *T. pusillus* is particularly interesting in showing this classic pattern very clearly, with the triploid replacing the diploid in the northern part of the range. In Britain the situation is particularly interesting (Sutton, 1966). The diploid form occupies all habitats in the extreme south-east of England, being gradually replaced by the triploid form in the north and west, first in open habitats and then in woodland. In transitional areas, as at Wytham in Berkshire, the triploid form occurs in the grassland and the diploid form in woodland. Although the two races intermingle they cannot breed, because parthenogenesis and the difference in chromosome number create an insuperable genetic barrier. Genetically speaking, these two forms of *T. pusillus* must be regarded as separate species, but they have not been given such taxonomic status because morphologically they do not differ sufficiently to merit it. They are generally treated as sub-species (see Chapter 8).

The precise factors determining the distribution of the diploid and triploid races are not known. The problem is a difficult one to study because there appears to be no convenient character on which to separate females of the two forms, so studies using mixed populations are not possible. A further puzzle is the presence of males in the parthenogenetic populations. They never exceed 5 per cent of the total number of adults but, on theoretical considerations, they would not be expected at all. It is known that they are triploid like the females but their origin and function remain a mystery.

5 FOOD, PREDATORS, AND PARASITES

This chapter is concerned with feeding or trophic relationships and considers woodlice both as consumers and as a source of food for a variety of other organisms. It is also concerned with the part that woodlice play in the transfer of energy through the ecosystems of which they form a part. Trophic relationships are summarized in fig. 14.

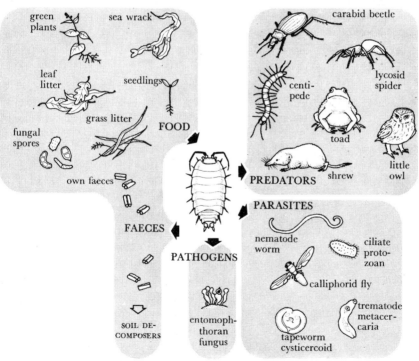

Fig. 14 The trophic relationships of woodlice; arrows indicate the flow of organic matter.

Food

To consider first the food woodlice eat, a simple general statement can be made: usually they feed on dead plant material. However, they will also feed, sometimes extensively, on animal remains and dung, while in some measure they may gain nourishment from living bacteria and fungi (Ing, 1967). Until recently, it was thought they never fed on living plant material (apart from seedlings) but it is now known that *A. vulgare*, in Californian grassland, acts as a grazer as well as a decomposer and is able to switch from living to dead material as resources change (Paris and Sikora, 1965).

Woodlice are usually regarded, along with millipedes and the earthworm *Lumbricus terrestris*, as being primary decomposers, the first link in the chain of organisms which breaks down dead plant material and mixes it with mineral particles to form soil. While it is true that these animals (and others like them) are responsible for the fragmentation of leaf litter, it has to be remembered that to whatever extent they feed on their own faeces (see p. 33) they have to be regarded as secondary decomposers. Furthermore, it is not at all clear that woodlice are primary decomposers in the sense that they are the first organisms to attack the litter. It may well be that a 'softening up' process has to be carried out by micro-organisms before woodlice will eat the litter, and it may be significant that leaves with a high concentration of tannins (which inhibit microbial activity) are left alone until these substances have been leached out by rain, whereas ash and sycamore leaves, which have a low tannin content, are acceptable much sooner after leaf-fall. This raises the problem of how good a food supply woodlice have in natural habitats. Looking at a thick carpet of litter it is easy to assume that food is abundant, but now that we know that until it has been on the ground some time much of this litter is unpalatable, the idea that woodlice live in the midst of plenty loses some of its plausibility. Quality, as well as quantity, would seem to be important. Much more information is needed at the moment on the palatability of different types of leaf at different times after leaf-fall, particularly in the case of grasses, which have never been tested at all.

It must not be thought that all the material ingested is absorbed as food through the gut wall. In fact, anything between 30 and 90 per cent by weight of the 'food' taken in is passed out as faeces, giving figures for assimilation from as high as 70 per cent to as low as 10 per cent. According to Hubble *et al.* (1965), when food is abundant it passes through the gut rapidly and there is little assimilation, but in times of scarcity ingested material is retained in the gut and much of it is absorbed.

This suggests that, for woodlice, the digestive approach of eating a lot and assimilating a little uses up less energy than doing things the other way round. Clearly the rate of passage of leaf litter through the gut will depend very much on the available food supply and where food is abundant woodlice can process large quantities, consuming about 3 per cent of their body weight each day.

The study of the role of woodlice in soil formation is not very easy but there are several lines that can be tackled without apparatus, such as studying litter-breakdown in cultures to which woodlice can be added or excluded.

The food assimilated in the gut provides energy for building up body tissue and for maintaining the life processes according to the equation:

$$A = P + R$$

where A is the energy assimilated from the food, R is the energy used up in respiration (a measure of the amount used in metabolism), and P is the energy bound up in body tissue or reproductive products, all expressed in calories. (These terms are from Petrusewicz and Mac-fayden, 1970). This equation sums up the energy flow through the individual or the population and is known as an *energy budget*. Its calculation allows comparisons to be made between groups of animals utilizing the same sources of energy (for example, primary decomposers). Thus, if energy budgets for populations of woodlice, millipedes, and earthworms are constructed, their relative importance in channelling the energy available in the leaf litter can be assessed (providing the population data are adequate). Further, the estimation of body tissue production indicates how much energy is available to those animals which use woodlice as food.

Assimilated energy is usually calculated as the difference between the energy content of the food and of the faeces derived from that food. These are found by measuring the heat produced when samples are burnt in a bomb calorimeter. (This is an apparatus in which the sample is ignited in pure oxygen at very high pressure so that all the combustible energy is released and can be measured.) Respiration rates are found by measuring oxygen uptake or carbon dioxide release from animals kept under controlled conditions of temperature and pressure. Body tissue production is measured in the same way as food and faeces values.

Using these methods, Saito (1965) found that more than 80 per cent of the energy assimilated by *Ligidium japonica* in Japan was used up in metabolism, which seems to be about the same level of efficiency shown by other macro-decomposers (earthworms, millipedes) although only a handful of figures is available.

To apply energy budget figures to field populations requires really comprehensive and accurate population statistics, together with a knowledge of how respiration and assimilation vary with the size of an animal, temperature, breeding condition, and time of year. Not surprisingly no such studies on woodlice have yet been completed but eventually this kind of approach should enable us to judge quite accurately the importance of woodlice in the soil and litter community.

The energy which is not used up in maintaining the life processes of the body can be exploited as food by other organisms which may be predators, pathogens, and parasites feeding on the living organisms or decomposers feeding on its dead remains. Little is known about organisms exploiting woodlice as a source of food and it is still impossible to make any estimate of the importance of predators and parasites of woodlice in energy transfer.

Predators

In the case of predators a paradox exists in that a great variety of animals are known to eat woodlice under laboratory conditions, but there is little evidence of sustained predation in the wild. In captivity, woodlice are eaten avidly by shrews and toads, and somewhat less enthusiastically by other vertebrates, including the little owl, hedgehogs, slow-worms, and frogs. Among the invertebrates which will feed on them in the laboratory are carabid and staphylinid bettles, spiders, harvestmen, and centipedes.

The lack of field records of predation could have several explanations. In the first place it might be simply that the unaccustomed conditions of captivity induce animals to eat woodlice—certainly one cannot *assume* that animals will eat the same foods in the wild as they do in confinement. If predation in the wild does occur to any extent, one must find reasons why it has so seldom been recorded. One fact that is certainly important is that the nocturnal and retiring habits of woodlice makes visual observation of predation very difficult and so indirect methods of study have to be used. The favourite method is visual analysis of the gut contents of suspected predators. Recently this has shown that woodlice can be important in the diet of shrews (Rudge, 1968). The drawback to this method is that woodlice, when eaten by shrews, are broken up into tiny fragments so that it is very easy to overlook any remains present.

A better approach, and one which can also be used on suspected predators which take in only liquid food (such as spiders) is to identify soluble remains rather than skeletal fragments of the prey. Two methods have been tried recently. The first is a radiotracer method in which prey are tagged with a radio-isotope and the isotope detected in predators. The drawback with this method is the need to disturb the prey in the tagging process, but it has been used with some success by Paris and Sikora (1967) who found that a type of ground cricket and lycosid spiders ate *A. vulgare*, although not in large numbers.

The second method makes use of the fact that it is possible to create antibodies which will react only with certain proteins specific to woodlice. Using such antibodies it is possible to detect these proteins wherever they occur, even in the guts of predators (until they are broken down by enzyme action). The test is very sensitive and has the advantage that the study area is not disturbed before the predators are removed for tests on their gut contents. Using this method (Sutton, 1970) I found that quite a wide range of predators on the Wytham Estate in Berkshire were eating woodlice, including lycosid spiders, centipedes, and shrews.

To use either test to estimate the number of prey removed from a population a great deal of further information is required. In the first place, one needs to know how many prey are represented by a single positive result—in the case of some spiders, only a single woodlouse is taken at each meal, whereas a shrew may eat 20 adult *Philoscia* at one sitting. Laboratory experiments on the feeding behaviour of suspected predators can give valuable help here, and can be extended to determine the extent of carrion feeding in each species. (Both tests give positive results with carrion.) Other necessary information is the persistence of prey protein or tracer in the gut (the time can vary from several days for spiders and centipedes to an hour or two in shrews) and the density of both predator and prey in the field.

It can be seen that estimating the intensity of predation is a very considerable problem, which accounts for our ignorance in this field. Much can be done in the way of predation studies by thorough observation of the size of prey preferred by different predators, consumption rates, and the palatability of the different species.

One predator which has been studied is a remarkable spider called *Dysdera*, a genus which has two British representatives which are quite plentiful in southern England. *Dysdera* is a formidable animal with jaws especially adapted for seizing a woodlouse in a pincer-like grip, while injecting a poison of such toxicity that the victim dies within as little as 7 seconds (fig. 15). Much has been made of the tegumental glands (p. 23) in woodlice as a means of defence against spiders, as a result of Gorvett's work (1956) showing that extracts of the glands are distasteful to some kinds of spider. Gorvett quotes experiments by W. S. Bristowe showing that the most distasteful species are those

57

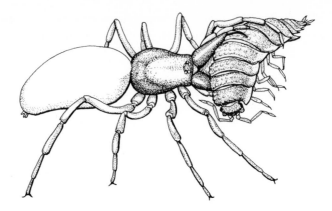

Fig. 15 *Dysdera* seizing a woodlouse; it injects venom through its fangs to paralyse the prey. The spider is 10 mm long.

with the best-developed glands. On the other hand, the glands obviously provide no protection against *Dysdera*, and lycosids will feed on woodlice in the laboratory. Unfortunately the latter group, although the most likely predators in most habitats, were not tested by Gorvett or Bristowe. The answer is probably that the degree of protection given by the glands depends on the hunger of the predator. One possibility that needs to be explored is that the glands provide less protection in very young animals where, in fact, mortality is heaviest.

Before leaving predation I must mention cannibalism, which is such a feature of cultures. Nearly all the victims are moulting animals which are vulnerable because they are immobile and their bodies are soft. It is natural to wonder about the importance of cannibalism in the field, but there is very good reason to think that, in nature, it is not important. In the field, an animal has freedom of movement and can isolate itself from the rest of the population when about to moult; in a laboratory culture, such escape is not possible and, as the animals are usually kept at high densities, cannibalism is likely to occur.

Parasites

Like predation, parasitism was long ignored as an important element in woodlouse ecology because no common parasites could be found.

Although they are intermediate hosts for a few trematodes and nematodes, woodlice have remarkably few parasites (Thompson, 1934). It is true that many nematodes are to be found in isopod corpses but these are thought to be the consequence, not the cause, of death because so few are found in living animals. The only group of insects which parasitize them belong to the sub-family Rhinophorinae of the Calliphoridae (bluebottles and blow-flies). Thompson could find no evidence of extensive parasitism by any of this group, but fortunately much more is now known about the flies through the work of Dr Robin Bedding (1965), to whom I am grateful for the following information.

Like most fly parasites, the 7 species of Rhinophorinae associated with British woodlice are parasitic in the larval stage and free-flying as adults. All are host-specific but occasionally occur in species other than the normal host. For some reason unknown, *Porcellio scaber* is much more heavily parasitized than other equally common and gregarious species like *Oniscus asellus*, while *Philoscia muscorum* is not attacked at all (see the accompanying table).

Woodlouse host	Number dissected	Parasite	Number found
Porcellio scaber	17055	*Parafeburia maculata*	1653
		Styloneuria discrepans	411
		Melanophora roralis	218
		Rhinophora lepida	90
		Frauenfeldia rubricosa	39
		Phyto melanocephala	2
Oniscus asellus	2677	*Parafeburia maculata*	14
		Styloneuria discrepans	11
Armadillidium vulgare	1758	*Phyto melanocephala*	35
Trachelipus rathkei	117	*Stevenia atramentaria*	9
Metoponorthus pruinosus	516		
Philoscia muscorum	383		
Ligia oceanica	212		
Metoponorthus cingendus	187	no parasites	0
Armadillidium depressum	100		
Cylisticus convexus	53		
Porcellio laevis	26		

Although Bedding found that 5 species of flies regularly attacked *P. scaber*, there were distinct differences in their habitat preferences and size of host preferred. *Melanophora* was characteristic of the upper sea shore; *Styloneuria* was associated with the more artificial habitats such as rubbish dumps and gardens; *Parafeburia* (formerly known as *Plesina*; fig. 16) was mostly found in colonies under loose bark: *Rhinophora* and *Frauenfeldia* preferred small animals. Thus, competition between the parasite species is less than appears at first sight.

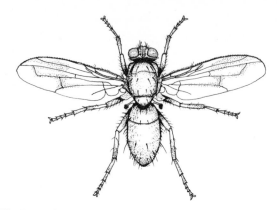

Fig. 16 *Parafeburia maculata*, the commonest fly parasite of woodlice, is about 6 mm long. Its usual host is *Porcellio scaber*.

The percentage parasitism of *P. scaber* populations varied a great deal, many small populations having no parasites at all, while in some of the larger ones more than 30 per cent of the individuals were infected. The degree of parasitism was influenced by the access of the colony to the female parasites, which need to get close to the resting colony to lay their eggs. They become photo-negative after copulation, crawl into the shelters, and lay their eggs on surfaces contaminated by secretions of the uropod glands. On hatching, the larvae take up a vertical attitude and attach themselves to any woodlouse which comes close enough.

Entry into the body cavity is made through an intersegmental membrane and the speed of penetration depends upon how recently the host has moulted. *Stevenia* is more remarkable than most in that it is apparently able to detect animals about to moult because it attaches itself to animals with 'white patches' in preference to others. Although a number of larvae may attach to the same host, one eliminates the others as a rule—a very necessary process if the survivor is to have enough food to complete its development. Within the body cavity the maggot at first feeds on the haemolymph but later attacks the gonads and the vital organs. The death of the host is followed by pupation of the replete parasite in the now-empty shell.

Adult flies of some species were found by Dr Bedding to be quite common in places. *Rhinophora* and *Styloneuria* occuring on flowers and *Melanophora* resting on walls and rocks. Other members of the Rhino-phorinae—*Morinia nana*, for example—have so far been found only as adults, their larval hosts (which may well be isopods) being unknown.

This work on fly parasites gives the first evidence of heavy parasitic attacks on woodlice and opens up several new lines of enquiry. The further study of the response of the female parasite to odours of the host would be rewarding, and more information is needed particularly on the population dynamics of host and parasite populations, with samples taken over a long period of time to determine the effect of the parasite on host numbers.

One group of organisms which have not yet been mentioned are the pathogens, which for our purposes may be considered to be microscopic parasites. They include protozoa, bacteria, viruses, and fungi. Little is known about the importance of them to woodlice. Epidemic disease does occur in cultures but the culprits have not been identified. Very few protozoa have been found within the body cavity—several occur on the pleopods, but these are commensals rather than parasites— and bacteria and viruses have been little studied. Pathogenic fungi seem remarkably rare, although recently Paris has found that one type is causing heavy mortality in populations of *Armadillidium vulgare* in California (see p. 68).

In this chapter we have been concerned with the relations of woodlice with other organisms and the account would not be complete without considering the impact of woodlice on man. To what extent are they beneficial to man and what is their status as pests? Any living thing active in the degradation of dead plant material must be considered beneficial because soil formation and nutrient cycling depend upon them. To this extent they are beneficial. They do not transmit diseases to man or to his domesticated animals and only rarely act as intermediate hosts for damaging parasites. This is fortunate (and perhaps surprising) in view of their close association with man. Woodlice can be looked upon therefore, with favour, as amiable companions rather than as a threat to man. Their only lapse from grace is an unfortunate weakness for nutritious young seedlings, which has led them at times to become a nuisance in greenhouses (Hussey *et al.*, 1969). Fortunately they are rapidly killed by modern insecticides and have not, so far, evolved resistant forms.

6 POPULATION ECOLOGY

To understand the ways in which woodlice numbers are limited or the significance of woodlice in communities and ecosystems we must consider such basic characteristics of populations as density, biomass, and the relative number of adults to juveniles. These are determined largely by rates of birth, growth, death, and migration, all of which, in turn, are influenced by such factors as food, enemies, and weather.

Recruitment

The knowledge of numbers recruited to the population through breeding is an essential statistic in any analysis of population dynamics. In many species, such as *Philoscia muscorum* and *Armadillidium vulgare,* breeding is highly synchronized, whereas in others this is not so. For many species, the British summer is long enough to raise only one brood although in the south-west some species, such as *P. muscorum*, do produce two. Triploid *T. pusillus*, being better adapted than other woodlice to cool climates, has two broods even in the north of England. The pattern and extent of the breeding season varies a good deal from year to year, largely depending on the weather. In the population of *T. pusillus* which I studied in grassland on the Wytham Estate in Berkshire (Sutton, 1968), severe drought during one summer so restricted growth of the juveniles that they failed to mature in time to produce the spring brood of the following year. In other words, virtually no young were born at the usual time. However, the population showed remarkable powers of recovery, because mortality of the retarded animals was low and a high proportion survived to produce young later in the year. Thus, by the end of the season, recruitment was not a great deal less than in the year before.

Working in California on *A. vulgare* Dr O. H. Paris found a very different situation. Here, the population does not seem to be able to compensate so quickly for setbacks to breeding and the numbers recruited each year vary tremendously.

Estimating the recruitment in a natural population using sampling

63

methods is a difficult business because some of the necessary information is hard to obtain. In the first place one needs an accurate estimate of the number of females which are going to produce young. One also needs to know the size of these females since the number of eggs carried increases with the size of the mother. Then, since the recruitment to the population is usually taken as the number set free from the brood pouches, one must estimate both the mortality of eggs in the brood pouch (something under 10 per cent in most species) and the number of pregnant females dying before they can set free their broods. Finally, one must allow for the fact that if the development time of the young in the brood pouch is short in relation to the sampling interval some broods will be missed and recruitment underestimated. The ideal is to obtain one accurate sample at a time when all the females which are going to breed are doing so.

Growth

The growth of individuals is another vital statistic. As with breeding, growth is confined to the warmer months of the year in this country and, above a threshold level reached in March or April, is proportional to the temperature of the surroundings. Growth is affected also by food supply, as for instance in the case of *T. pusillus* in Wytham, where the summer drought of 1964 dried out the litter, forcing the animals to migrate deep into the soil to avoid desiccation. As a result, growth ceased because of lack of food (fig. 17). The drought ended in October and a small amount of growth of the juvenile group then took place. This was quickly stopped by the onset of winter. In the following year,

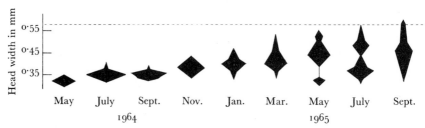

Fig. 17 Kite diagram to show the rate of growth of young *Trichoniscus pusillus pusillus*. The vertical axis gives a measure of the size of the animals and the shapes of the kites indicate the relative numbers of animals of different sizes. The dotted line indicates adult size. In 1964 there was no growth between August and October because of a drought. In 1965 there was no drought between these months and rapid growth of the equivalent age group occurred.

the weather was normal and the litter damp enough to provide food all summer. Thus, growth of the juvenile group was continuous.

Because growth can occur only at the moult when the animal's cuticle is soft, it proceeds by fits and starts in any one individual. Moults occur at frequent intervals in young animals and continue until the animal dies, but the periods between moults become exceedingly long and the growth-rate correspondingly slow. During the breeding season female growth stops because they cannot moult while carrying a brood: (moulting involves shedding the brood pouch). But in spite of this handicap it appears that, in many species, males grow more slowly than females at all stages of development, this certainly being true in *T. pusillus*. There are, of course, advantages in having females of as large a size as possible, because the number of young produced is linked to size, whereas the effectiveness with which a male carries out its reproductive duties may depend more upon its activity and agility than its size. Such considerations may also explain the low ratio of males to females found in many woodlouse populations, because one active male may be able to fertilize a number of females.

A striking characteristic of isopod growth is the great range of individual growth rates (Sutton, 1970). This is least evident in very young animals which all grow very rapidly, but the variation in older animals is so great that the slowest growing progeny of one year are frequently overtaken by the fastest growing individuals of the next. In populations where this overlapping is extensive it can be difficult to trace the mortality pattern of each group of young from birth to death, although it is only towards the end of their life-span that difficulties become acute. One consequence of the variation in growth rate is that the size of an individual is a very poor guide to its age.

Mortality

All woodlice populations so far studied show the same mortality pattern, with the heaviest losses occurring in the newly liberated young and the lightest mortality in the oldest animals (fig. 18). Superimposed on this basic survivorship curve are a number of anomalies caused by irregular events such as floods and disease, and sometimes by seasonal effects.

The losses on release from the brood pouch are in striking contrast to the low mortality of the stages developing within it. What it is that kills off these young animals in such quantity is not known but, being small, they are very prone to desiccation. Also, it may well be that their behavioural responses are not as well developed as in older animals and are less effective in protecting them from dangerous conditions. It

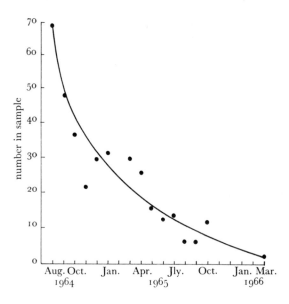

Fig. 18 Mortality pattern of *Philoscia muscorum*. The graph shows the decline in numbers of a group born in one season. Mortality is heaviest in the youngest stages. Each point represents a sample from the same population; deviations from the curve are caused by sampling errors.

would be very interesting to know if very young woodlice react to adverse conditions in choice-chamber experiments as speedily and as surely as do their parents. Other possible explanations of the high mortality are that their defences against predation are less well developed or that, being much smaller, they are exposed to a wider range of predators than the adults.

As for mortality in later life, it may be that predation and parasitism are important but so far there is little detailed evidence. While it is known that climatic excesses such as floods can cause heavy mortality, we also know that other bad conditions like frost and drought are avoided by vertical migration into the soil where the animals can survive for weeks on end without 'fresh' food. Perhaps they survive by recycling their faeces, as discussed in Chapter 2.

Migration
Lateral migration along the ground does not seem to be a common

66

feature of isopods, in contrast to millipedes, where such migrations are both frequent and spectacular. Although marked individuals have been shown to move quite long distances in a night (up to 13 metres), there seems little tendency for concerted movements in one direction to take place, although Stachurski (1968), working on a population of *Ligidium hypnorum* in a Polish swamp, did find movement in and out of the wetter areas according to season. A check on lateral migration can be made by putting out pitfall traps (see page 106) which will indicate the degree of surface activity of a particular species, if not direction of movement. Otherwise, it can be spotted only by a close watch on changes in population size, structure, and density.

Vertical migration seems to be an integral feature of many isopod populations, occurring as a response to unfavourable conditions in the litter layer, as noted earlier, in the case of the *T. pusillus* populations at Wytham. There are also the migrations, frequently noted, of *Porcellio scaber* from the litter layer to tree trunks where the animals shelter in bark crevices during the summer months, coming out at night to graze on the green film of single-celled algae (pleurococcoids) growing on the bark. Vertical movements may be irregular, to avoid occasional bad conditions, or seasonal, like the movements of *A. vulgare* in California (Paris, 1963), or the migrations of *P. scaber*. Movement of the whole population is characteristic of these vertical migrations and their occurrence is fairly obvious, except that migration below sampling depth in the soil can be mistaken for mass mortality.

Population stability

Populations of woodlice, like those of other organisms, are influenced by a variety of factors. The type of factor influencing population size is said to be density-dependent if, as the density of the population rises, its effect becomes more severe; that is, a higher proportion of the population is killed, fails to breed, or emigrates. Eventually a point is reached where losses cancel out recruitment and population growth is halted, while if there is any time-lag in the operation of a density-dependent factor, losses will for a time exceed recruitment and numbers will fall. However, sooner or later, the declining effect of the factor will allow the population to increase again and the cycle is repeated. The result is that population density see-saws about an equilibrium at which losses and recruitment are in balance and by creating this balance, density-dependent factors regulate or stabilize the population. The actual level of equilibrium usually varies a great deal from time to time because a habitat is never wholly stable and the number of animals it can support keeps on changing. The complexity of changes

67

in population size is even further increased by the action of factors which are density-independent; that is, the proportion killed, failing to breed, or emigrating is not related to density. Such factors cannot bring a population into equilibrium but they can have radical and irregular effects on numbers. So far as woodlice are concerned there appear to be very few effects which can be regarded as completely density-independent. The usual examples given for other animals— the effects of frost or drought, for example—will probably act in a density-dependent way on woodlice because, in most habitats, they can find shelter only in a limited number of sites, so that the proportion of the population gaining protection and surviving will depend on density. Thus, in the case discussed by Paris (1963) where *A. vulgare* suffered heavy mortality during floods, an element of density-dependence were present because shelter sites were limited in number and, at high densities, mortality through drowning would be proportionately greater than at low ones.

While it is reasonable to expect that some forms of predation, parasitism, and pathogenicity operate in a density-dependent manner on woodlice, as they often do on other organisms, there is as yet little evidence to support such an assumption or to identify which factors may be important.

Indeed, Clark (1970), in a detailed study of population stability in *A. vulgare*, found that adult toads (*Bufo valliceps*) acting as predators in experimental field enclosures had no density-dependent effect on the prey. However, changes in the density and structure of his pillbug populations did strongly suggest that density-dependent factors (which he could not identify) were stabilizing the population by influencing recruitment and winter survival.

Cannibalism, which would be expected to act in a stabilizing manner, is probably not an important factor in natural populations, as explained on p. 58. A factor which does appear to have a density-dependent effect has been discovered recently by Paris in California. He has found that a major cause of mortality at high densities is a pathogenic fungus of the genus *Entomophthora* (well known as pathogens of insects). This appears to have caused a spectacular crash in numbers from 5000 per square metre (m^2) in late 1968 to $150/m^2$ the following spring, and its action may well prove to be density-dependent. The pathogen is interesting not only because of its capacity to cause mass mortality but because pathogenic fungi are otherwise almost unknown in woodlice. It would be interesting to know if it was carried over to the United States by the pillbug or whether it was caught from some native species in California.

Density

Woodlouse densities typically show a sharp increase on release of the young, followed by a rapid decline as the majority of these young meet an early end. This decline tails off to give rather stable numbers just before the new brood is set free.

Densities of woodlice are usually expressed in numbers per square metre and may be very high. Maximum densities for *Philoscia* in Britain just before the liberation of young are $250/m^2$; for *A. vulgare* in Californian grassland $5000/m^2$ (with a low of $10/m^2$); and for *T. pusillus* densities of $1000/m^2$ are common in ungrazed grassland even outside the breeding season. The maximum ever recorded for any woodlouse is $7900/m^2$ for *T. pusillus* from scrub grassland in the foothills of the Chilterns. This count was made during the breeding season but even so a quarter of the animals were adult, so the high density was not a freak effect due to sampling at the moment when the young were released.

Biomass

The biomass of a population is the total weight of animals in it, measured either as live wet weight or dry weight. Biomass is a more useful statistic than density when comparing species of very different sizes, such as *A. vulgare* and *T. pusillus*. Density figures of $1000/m^2$ mean very different things for these two species because of their difference in size. Thus, a large adult *A. vulgare* is about 70 times as heavy as a large adult *T. pusillus*. Also, when considering the importance of a species as a decomposer or a source of food, biomass is more meaningful than density, particularly as energy content can be easily derived from it.

| | March 1964 | | March 1965 | |
	biomass (g/m^2 live wt.)	density (no./m^2)	biomass (g/m^2 live wt.)	density (no./m^2)
Trichoniscus pusillus	1·19	1490	0·30	1050
Philoscia muscorum	0·70	103	0·57	103
Armadillidium vulgare	0·21	6	0·16	6
Total woodlice (these 3 species) +*T. pygmaeus*	2·137	1820	1·045	1350

Biomass figures for the 3 commonest species of woodlice in the grassland of Wytham Woods (Sutton, 1966) are shown here as an

69

indication of the sort of values that probably can be expected for ungrazed grassland in this country. However, in very favourable habitats such as maritime grassland, a figure of $5 \, g/m^2$ seems quite possible. Biomass figures (like densities) are difficult to compare because they vary from year to year (as in the values quoted here) and from season to season, but used with care they can provide valuable information.

Apart from comparing woodlouse populations in different sites, the figures can be used as an indication of the importance of isopods relative to other groups of soil and litter animals in particular habitats. In ungrazed grassland mites are always vastly more abundant with densities of over $200\,000/m^2$ recorded. Because of their much smaller size, however, their biomass is about the same as that for woodlice. The same is true of Collembola. In contrast, lumbricid earthworms have a lower density than woodlice but a much higher biomass, reaching levels of $100 \, g/m^2$. This is a reflection of their large individual size. Woodlouse estimates compare most closely with those for millipedes and insect larvae, but tend to be somewhat lower than either (figures in Wallwork, 1970). Density and biomass in woodland, judging from published figures and some of my own, will always be very much lower than those in ungrazed grassland, which may reflect the greater amount of shelter in rough grassland and perhaps a greater degree of plant productivity and turnover.

Conclusion

The study of the population dynamics of woodlice really has two aims: 1, to discover how populations are stabilized; 2, to provide information on which to base estimates of their importance to the soil/litter communities in which they occur. Population stability has already been discussed but it is worth considering the second aim a little further. The main problem is to decide what makes a group important (amount of food eaten, role in soil formation, availability as food) and then to find convenient but accurate ways of measuring these. Density and biomass have long been used as yard-sticks of influence but unfortunately neither gives much indication of energy flow or importance in soil formation. A satisfactory and convenient measure of importance in soil formation has yet to be devised. However, there are hopeful signs that respiration can be used as a guide to energy flow, although ideally to estimate the latter one needs to work out all three terms of the energy budget (see p. 55). Comparative respiration figures are being worked out for such groups as mites and earthworms but woodlice have hardly been tackled from this point of view, and so one has to fall back on density and biomass as measures of importance, however inadequate. Using these two criteria, it looks as though woodlice may be a significant element of some soil and litter communities, although rarely reaching the importance of earthworms and millipedes. These remarks apply to temperate regions. In most tropical regions, woodlice appear to be very scarce.

Further reading for Chapters 5 and 6:
Wallwork, J. A.   1970   Ecology of Soil Animals. McGraw-Hill, London.

7 DISTRIBUTION AND HABITAT RANGE

The environmental factors limiting the distribution of woodlice have not, in most cases, been studied in much detail, but we know enough about the biology of the animals to pinpoint those factors which are likely to be important.

Climate

Climate influences distribution in a multitude of ways but most of its effects—rainfall, for example—are indirect. Rainfall operates through its effect on food supply and soil conditions. Of the direct effects, summer heat and winter frost are probably the most important.

Britain is badly off for woodlice compared with southern and eastern Europe, mainly because the relative shortness and coolness of our summers prevent adequate growth and successful breeding of the more tender species. Even within Britain we have quite a few species confined to the south-east, such as *Trachelipus rathkei*, *Cylisticus convexus*, and *Ligidium hypnorum* (map, plate 8), and the warm summers for which this part of the country is noted is probably the reason. On a smaller scale some species, even in the south, only flourish on hot, south-facing hill-sides where heating by the sun is intense. The differences in species between the north and south-facing slopes of Cheddar Gorge illustrate this point very well. Lack of sun is probably also responsible for setting the upper limit for woodlice on British mountains. *T. pusillus* has been found at 800 m on Ben Lawers in Perthshire but most species are strictly lowland.

The fall in the number of species as one travels from east to west across the British Isles would be more striking if it were not for a small group of species like *Metoponorthus cingendus* which need mild winter weather and are found in the oceanic climate of western England (map, plate 8). The origin of this interesting group is discussed later.

A third group is made up of species which are quite at home in the British climate and occur at all latitudes from Cape Wrath southwards. In this group are those pillars of the woodlouse fauna *Oniscus asellus*, *Porcellio scaber*, and *Philoscia muscorum*. The parthenogenetic *Tricho-*

niscus pusillus pusillus deserves special mention because it is the only form centred on the north and west (map, plate 8). As discussed in Chapter 4, the greater breeding capacity of the parthenogenetic population may give it an advantage in the cooler parts of the country over its bisexual relative *T.p. provisorius*, which occurs only in the south and east (map, plate 8).

The final grouping of species determined by the direct effects of climate are those which benefit from the artificial climate created by man in his houses and heated greenhouses. A number of species on the British list are known only from Kew and other botanic gardens, while there are other species which have been able to extend their range in the north by colonizing greenhouses. This is probably how *Cylisticus convexus*, normally a denizen of chalk downs in the south, became established in the centre of Leeds, a rather different environment. Another artificial environment of great benefit to alien species is that created by large compost heaps which become 'centrally heated' by bacterial activity and are a haven for frost-sensitive woodlice in winter. Warm compost heaps may well have been the first outdoor foothold of introduced species which later became adapted to the British climate and can now survive and breed out of doors well away from human habitation. There is some reason to believe that *Porcellio laevis* and *Metoponorthus pruinosus* have become naturalized in this way.

Influence of the sea

Apart from allowing lime-loving species to exist on non-basic coastal rocks, the sea has a very strong influence on the distribution of several other species. *Ligia oceanica, Halophiloscia couchi* and *Armadillidium album* are all *halophilic* (salt-loving) species typical of the littoral zone. *Ligia* is thought to be primitive in this respect, but for the other two the shore has been reached through more terrestrial habitats. All three species are now so closely adapted to the shore environment that they are unable to survive more than a few metres from the water's edge, although *Ligia* is occasionally found more than a hundred metres above sea-level on cliffs exposed to salt spray, as on St Kilda. Which elements of the littoral environment actually determine the distribution of these three species is not known.

Three other species are characteristic of the coast. *Acaeroplastes melanurus* and *Eluma purpurascens* have only been found in the Dublin area, but the third, *Metoponorthus cingendus*, occurs in south-west England and also in Ireland (map, plate 8). All three have a western distribution suggesting that frost sensitivity may be responsible for their restriction to coastal habitats.

Soil pH and calcium content

The poverty of the woodlouse fauna on acid soil is well known and the assumption is always made that this is because such soils lack sufficient calcium to enable woodlice to build up their calcareous exoskeletons. While this argument seems eminently reasonable, it must be stressed that there is no direct evidence to support it. Furthermore, there are a few awkward facts to be explained away, such as the occurrence on the acid Surrey heaths of species like *Porcellio scaber* with quite heavy exoskeletons. The swarms of woodlice on non-basic rocks along the coast presumably obtain their calcium from salt spray and cast up sea-shells, but it would be interesting to know how far the inland penetration of these woodlice is governed by the presence of calcium. By and large, the heavier the exoskeleton of the species the more likely it is to be closely confined to calcareous soils. This is particularly true of the pillbugs, which have a massive exoskeleton, to the extent that they resemble miniature tanks. Since calcareous soils are mainly concentrated in the south-east of the country, the problem arises whether those species confined to the south-east are limited by the climate (as suggested previously) or by the lime content of the soil. The answer is revealed by looking at the fauna of the chalk wolds of Yorkshire, where none of the south-eastern species penetrate, even into the low-lying parts, so clearly, they are limited to the south-east primarily by climate.

Just as man creates artificial climates, so he creates artificial calcareous soil by using lime to make mortar and by building in limestone. *A. vulgare* and also rarer species like *Porcellio spinicornis* take advantage of such activities, and the greater density and variety of woodlice around old buildings reflects the general observation that calcareous soils favour the development of a rich isopod fauna.

Soil type and drainage

We have seen that, to develop a rich isopod fauna, a soil should have a good lime content. Such soils tend to be neutral or alkaline in pH, rich in other macro-decomposers such as earthworms and millipedes, and also support a flourishing bacterial population. The result is that plant remains are rapidly broken down, there is a good crumb-structure giving generous aeration, and thorough mixing of organic and mineral matter, all of which are characteristic of a *mull* soil. A mull soil is typified by a gradual transition between soil and litter layers and a crumbly texture. On non-calcareous acid soils, particularly where rainfall is high enough to cause waterlogging and hence oxygen-depletion, earthworms and other macro-decomposers are scarce, and bacteria are replaced by fungi. Plant breakdown is slow and a *mor* soil develops. A mor soil can be recognized by the abrupt transition between soil and litter layers and compacted nature of the former which results from the lack of air spaces. Between the extremes of mull and mor are a great number of transitional soils. There are also other soil-types which develop under special conditions, such as the thin rendzina soil of chalk downs or the sandy podsols of heathland.

Woodlice are characteristic of mull soils, which they help to form and maintain. In contrast, only *T. pusillus*, *O. asellus*, and *P. scaber* are likely to be found on mor soils, presumably the low calcium content and poor drainage keeping out the other species. The dependence of woodlice on good drainage stems, of course, from the inability to control water uptake and problems of oxygen uptake when under water. Few woodlice are found in areas liable to prolonged flooding, although *Ligidium hypnorum* is characteristic of marshy ground and *T. pusillus* also occurs in waterlogged habitats.

Food and shelter

Here we must consider shelter not only from climatic extremes, but from predators and parasites, as well as from destruction of the habitat by ploughing or, in pasture, trampling by livestock. Taking the question of shelter in relation to climate, the first thing to say is that the importance of shelter in limiting distribution is difficult to separate from the importance of food, since the litter in which the majority of woodlice live seems to provide both. Obviously, woodlice will occur only where food is available—the question to ask is whether there are situations in which food is available but cannot be utilized for lack of shelter from climatic extremes. The presence of woodlice in a number of desert habitats as well as on mull soils where drought makes the litter uninhabitable in dry summers shows that woodlice can survive, provided the litter is periodically moist enough to be eaten, and refuge is possible in crevices below ground. Shelter sites are necessary but they do not need to be in the food, and as long as some form of shelter is available nearby the food will be eaten. One source of food that is not utilized by woodlice is the encrustation of single-celled green algae on smooth tree trunks and well kept brick walls which have no crevices for the animal to shelter in.

We should not forget that shelter from disturbance and from enemies may also be as important as shelter from climatic extremes. Gross disturbances by ploughing probably account for the virtual absence of woodlice from arable land, while the tendency of *Philoscia* to form aggregations in tussocks of *Dactylis glomerata* (cocksfoot grass) could be because of the protection against predators provided by the close-set stems.

Associations with other animals

The only species which is limited in its distribution in this way is *Platyarthrus hoffmannseggi*, which is a commensal of ants and is almost entirely confined to their nests, although it is occasionally found in the tunnels of wood-boring beetles as well. It probably lives on the excreta of its hosts and is particularly fond of the nests of *Lasius flavus*, the yellow ant, except perhaps those on acid soils. *Porcellio scaber* is also often found in ants' nests, but no other British species have associations of this kind. *Platyarthrus* can be very abundant in a nest and seems to suffer little hostility from its host, although Gorvett and Taylor have described how it will 'clamp down' and discharge its defence glands if molested (p. 47). This defensive reaction is evidently effective.

Large scale movement and the origin of the British fauna

Having examined the main factors determining range, it is time to look at the ways in which different species came to occupy their present ranges, with particular reference to the origin of the British fauna, the past history of which is closely related to the repeated advance and retreat of the ice-sheet across Britain during the last Ice Age. This ice sheet varied in extent but, at its maximum, covered the whole of Britain except for England south of the Thames, and was bordered by a large area of tundra to the south. The amount of water bound up in the ice-cap was so great that the ocean level was some hundreds of feet lower than it is now and a wide bridge extended between England and the Continent, while extensive areas of the Continental Shelf south-west of Ireland were also exposed. When the ice began its final retreat 10 000 years ago, plants and animals living to the south and east extended their range northwards across the dry bed of the channel into England and Ireland from which the arctic conditions had originally driven them. The most hardy of these invaders made their way right up to the north of Scotland, and among woodlice are typified by such species as *Philoscia muscorum*, *Oniscus asellus*, and *Trichoniscus pusillus pusillus*. Less hardy species crossed from the Continent later on but never penetrated very far to the north and west; these make up the south-eastern element of the fauna and flora. A good example is *Ligidium hypnorum* (map, plate 8). One of these late arrivals, *Trichoniscus pusillus provisorius*, appears to have ousted its close relative *T. pusillus pusillus* from south-east England but has not been able to penetrate further north than Yorkshire (see p. 51) although there is one very puzzling record for western Ireland. As the ice melted, the water rose until first Ireland was isolated and then the land bridge with the Continent was broken, isolating Britain from further invasion. It may be that there are species which could survive the present British climate but did not reach the land bridge in time to make the crossing. Certain species present on the French coast come to mind.

Another, and rather mysterious, group of species is that mentioned earlier as having a western distribution, occurring in Ireland and south-west England. This group is thought (Beirne, 1952) to have weathered the Ice Age on the exposed Continental Shelf south-west of Ireland and to have re-invaded Ireland from there. Nearly all the woodlice with this distribution are, in fact, members of the 'Lusitanian element'; that is to say, their distribution is centred on Spain and Portugal, with Britain being the northern limit of their range. (Lusitania was a province of the Roman Empire incorporating parts of Spain and most of Portugal.) The Lusitanian species are *Acaeroplastes melanurus, Eluma purpurascens, Metoponorthus cingendus*, and *Oritoniscus flavus*; they make up about 12 per cent of the indigenous British fauna, an intriguingly high proportion. Two other species with a western bias in their distribution, *Armadillidium depressum* and *Halophiloscia couchi*, are not Lusitanian because they do not occur in south-west Europe. However, they probably also invaded Britain from the western seaboard.

The fourth element of the fauna is made up of species which have been brought to this country by man accidentally. Some woodlice have become *synanthropes*—that is to say they flourish in man-made habitats such as cellars, gardens, and greenhouses, from which they are inadvertently transported with potted plants and such like to all corners of the globe. The attention needed to keep the plants alive ensures the survival of the woodlice. Introductions have also resulted from the practice of taking on earth as ballast on sailing ships. Lindroth (1957) has shown how the carabid beetle fauna found around the port of Halifax in Nova Scotia owes a lot to ballast brought from Poole Harbour in Dorset, and woodlice have been carried about in the same way (Palmén, 1958).

On balance we may have exported more than we have imported, because common British species such as *Porcellio scaber* are now found all over the world. They might, of course, have come from European ports. In the United States some of these imported species have flourished exceedingly, notably *Armadillidium vulgare* in California and *Trachelipus rathkei* in the north-east. They seem either to have taken up unfilled niches or to have been able to exploit the resources more successfully than indigenous competitors because both are now influential members of their respective communities.

On the import side we have a number of species — *Trichorhina tomentosa* from tropical America and *Agabiformius lentus* from the eastern Mediterranean — which are found only in heated greenhouses and are unable to survive the British winter out of doors. These I shall call aliens. Then there are others which were originally introduced but have now become naturalized and can survive the British winter out of doors. These I shall term naturalized species. *Trachelipus ratzeburgi* is a prime suspect here, if indeed it is British at all. Recent efforts to find British specimens of this species have failed and although there is still much material to be checked it seems quite possible that past records really refer to its close relative *T. rathkei*, for which it can easily be mistaken (see p. 97). The continental distribution of *T. ratzeburgi* strongly suggests that if it has occurred in Britain at all, it has never been native, because it is completely absent from the English Channel seaboard and has a central European distribution. Any westward spread from this area to Britain (unaided by man) would surely have involved northern France and the Low Countries.

One outstanding puzzle about the distribution of woodlice in Britain concerns the apparent scarcity of many species such as *Trichoniscoides albidus* and *Haplophthalmus danicus*, previously recorded over a wide area but now seldom seen. Most of the older records were made at the turn of the century, when hunting woodlice enjoyed a considerable vogue among the clergy. Undoubtedly the low level of current interest in the group is partly responsible for the scarcity of recent records, but misidentification and a previous tendency to do one's collecting in the Rectory greenhouse may also be part of the explanation. Nevertheless, the possibility of a dramatic restriction in range and abundance over the past 60 years cannot be excluded by any means.

I would like to end this chapter on a practical note with a mention of the oniscomorph millipede *Glomeris marginata*, which is very common in Britain and can be mistaken easily for a pillbug, thus causing much confusion in the recording of woodlouse distributions. Except when very small, *Glomeris* has more pairs of legs than a woodlouse (which has seven pairs) and this is the easiest way of spotting it. In colour it is nearly always a uniform dark brown or black, with white margins to the dorsal plates. There are no uropods and the antennae are club-shaped rather than tapered as they are in woodlice. In mature males the last pair of legs is heavily built and is used for transferring sperm to the female. *Glomeris* is widely distributed in the British Isles

79

and its habitat range overlaps that of the commoner pillbugs very considerably, so the opportunities for confusion are legion.

8 IDENTIFICATION OF BRITISH WOODLICE

The following pages are devoted to the practical matter of identifying the British species of woodlice, and giving brief notes on their ecology and distribution. Keys to families, genera, and species are provided and, where difficulties are likely, illustrations of the key characters used are also given. Most of the bigger woodlice can, with experience, be recognized at a glance because of their colour patterns and general appearance; but it is very difficult to construct an easy formal key because so many of the more obvious characters (such as colour) vary so much in each species.

The usual drawback of keys to woodlice is that the genitalia and mouthparts have to be dissected—a delicate job at the best of times. To avoid this problem, this key is so arranged that all the commoner species can be identified using easily seen features which require no dissection. To separate the rarer species we have had to fall back on genitalia in some cases, however. These rare species are not included in the family and genus keys. Instead, they are treated after the common species with which they key out. Thus, *Oritoniscus flavus* and various *Trichoniscoides* species will be found to key out under *Androniscus dentiger*. Note that the millipede *Glomeris* looks very like a pill-bug at first glance. However, it has clubbed antennae and more pairs of legs (see p. 79).

We have tried to use features which will identify juveniles as well as adults wherever possible, and have also tried to use features which can be seen both in fresh and preserved specimens. However, we have had to make use of the number of pseudotracheae present to separate the genera *Trachelipus* and *Porcellio*, although these organs are difficult to see in alcohol. You should also note that alcohol causes many colours to fade. Therefore, wherever possible, the colour and pseudotracheal characteristics should be noted before preserving the animals.

The keys provided here are intended to complement, rather than to replace, Edney's key to the British woodlice (1953) which gives a much fuller description of each species and many useful diagrams. However, it has not proved to be very easy for beginners to use and some of the key characters, notably the shape of the head in the genus *Porcellio*, vary too much to be reliable.

The first comprehensive key to the British woodlice, by Webb and Sillem (1906), is now really of historical interest only, but is well worth looking at for the illustrations alone. Other keys will be found in Vandel (1960 and 1962; in French) and in Gruner (1966; in German); these refer to the French and German faunas respectively. Between them they include all known British species, native or otherwise. In both countries the number of species involved is far greater than in our case and the keys are correspondingly more complicated. Both works contain a great deal of information on ecology and distribution and are essential fare for the specialist.

In recording the geographical distribution of each species we have distinguished between records made before, and after, the publication of Edney's key in 1953, because we feel that before this date the means of identification available in English were inadequate and know that a number of the older records are incorrect. Furthermore, the old records often omit details of habitat, so the status of species as native, naturalized, or alien in various parts of the country cannot be assessed. In putting the earlier records on one side we have undoubtedly understated the true distribution of several rare species but, on the other hand, we hope we have created a secure base for future work.

In the species entries the generic name is given first followed by the specific name; then follows the name of the author who first described the species, and then the date of publication of this description. If the species is now placed in a different genus from that used in the original description, the original author's name appears in brackets. After the date of description comes an estimate of the usual maximum length of the species, from the front of the head to the tip of the telson. The heading ends with a list of the illustrations in which the species is figured. Each entry for non-alien species ends with the number of vice-counties from which the species has been recorded in England, Wales, Scotland, and Ireland since 1953. Thus, if a species was recorded in every single vice-county in the British Isles the entry would read $E59W12S41I40$. For details of the vice-county boundaries in England, Wales and Scotland, consult publication no. 146 of the Ray Society (1969).

The check list gives all the species which have been recorded from Britain. Only 29 of the 42 species listed can, with any confidence, be regarded as native; the rest have all been accidentally introduced by man. Some of these have become naturalized but the others are aliens (p. 79).

Check list of British species of woodlice

Sub-order Oniscoidea

Series Ligienne

Family Ligiidae

Ligia oceanica (Linnaeus)
Ligidium hypnorum (Cuvier)

Family Squamiferidae

Trichorhina tomentosa (Budde-Lund)†
Platyarthrus hoffmannseggi Brandt

Family Oniscidae

Halophiloscia couchi (Kinahan)
Chaetophiloscia meeusei Holthuis †
Chaetophiloscia patiencei (Bagnall) †
Philoscia muscorum (Scopoli)
Oniscus asellus Linnaeus

Family Cylisticidae

Cylisticus convexus (De Geer)

Family Porcellionidae

Trachelipus rathkei (Brandt)
Trachelipus ratzeburgi (Brandt)
Nagurus cristatus (Dollfus) †
Nagurus nanus Budde-Lund †
Metoponorthus cingendus (Kinahan)
Metoponorthus pruinosus (Brandt)
Acaeroplastes melanurus (Budde-Lund)
Agabiformius lentus (Budde-Lund) †
Porcellio laevis Latreille
Porcellio dilatatus Brandt
Porcellio spinicornis Say
Porcellio scaber Latreille

Family Armadillidiidae

Eluma purpurascens Budde-Lund
Armadillidium nasatum Budde-Lund
Armadillidium depressum Brandt
Armadillidium vulgare (Latreille)
Armadillidium pictum Brandt
Armadillidium pulchellum (Zencker)
Armadillidium album Dollfus
Reductoniscus costulatus Kessleýak †

Series Trichoniscienne

Family Trichoniscidae

Androniscus dentiger Verhoeff
Oritoniscus flavus (Budde-Lund)
Trichoniscoides albidus (Budde-Lund)
Trichoniscoides saeroeensis Lohmander
Trichoniscoides sarsi (Patience)
Trichoniscus pusillus (Brandt)
Trichoniscus pygmaeus Sars
Cordioniscus spinosus (Patience) †
Cordioniscus stebbingi (Patience) †
Miktoniscus linearis (Patience) †
Haplophthalmus danicus Budde-Lund
Haplophthalmus mengei (Zaddach)

The third oniscoid group, series Tylienne (see Introduction) has no British representatives.

† alien species

Adapted from Vandel (1960, 1962, 1965)

1 Flagellum with 10 or more bead-like sections
(fig. 19A); endodite of uropod joined to
basipodite well beyond tip of telson (fig. 19E)   Ligiidae p. 86
Flagellum with indistinct sections tapering to
fine brush of bristles (fig. 19B); endopodite
of uropod joined to basipodite level with tip
of telson (fig. 19F)   Trichoniscidae p. 87
Flagellum with 2 or 3 distinct sections and
rarely more than 3 bristles at tip (fig. 19C,D);
endopodite of uropod fused to basipodite
beneath telson (Fig. 19G,H,I)   2

2 Flagellum with 3 sections (fig. 19C)   Oniscidae p. 93
Flagellum with 2 sections (fig. 19D)   3

3 Eyes absent*; body almost pure white and
very stout   Squamiferidae p. 95
Eyes present; animal never pure white and
body more or less elongate   4

4 Exopodites of uropods flattened and spear-
shaped (except in *Metoponorthus*), projecting
well beyond tip of telson (fig. 19G);
flattened animals, slope of epimera as in
fig. 19J†; cannot roll up into a ball   Porcellionidae p. 96
Exopodites of uropods tubular and elongate,
projecting well beyond tip of telson (fig. 19H);
somewhat arched animals, slope of epimera as in
fig. 19K; can roll into ball but seldom does so   Cylisticidae p. 100
Exopodites of uropods four-sided and plate-
like, projecting only slightly beyond tip of
telson (fig. 19I); strongly arched animals,
slope of epimera as in fig. 19L; can roll into
a ball   Armadillidiidae p. 101

*except the rare white alien *Trichorhina tomentosa* which usually
has eyes, each of a single ocellus
†the rare alien *Agabiformius lentus* is arched, but has spear-shaped
exopodites. In *Metoponorthus* the uropod exopodites are tubular,
but the animals are very much flattened

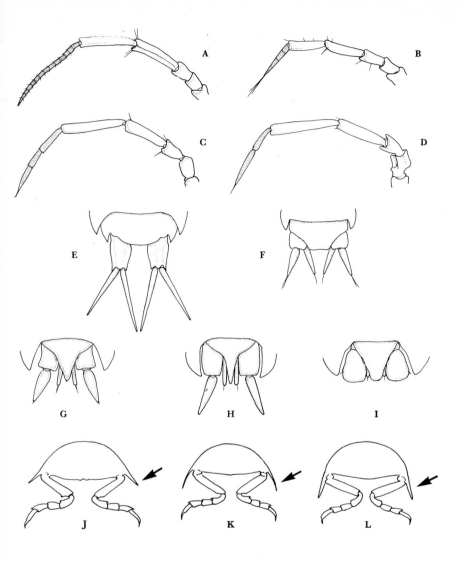

Fig. 19 Dorsal view of the left antenna (flagellum shaded) of **A** *L. hypnorum;*
B *T. pusillus;* **C** *P. muscorum;* **D** *T. rathkei.*

Dorsal view of uropods and telson of **E** *L. oceanica;* **F** *H. mengei;* **G** *P. scaber;*
H *C. convexus;* **I** *A. nasatum.*

T.S. of pereion to show outline and epimera slope in **J** *P. scaber;* **K** *C. convexus;* **L** *A. vulgare.*

FAMILY Ligiidae
key to genera

1 <u>Endopodite</u> and exopodite of uropod joined to
basipodite at same place; outline of pereion
and pleon continuous (fig. 20A); littoral
habitats only   *Ligia*
<u>Endopodite</u> fused to the long projection of
basipodite; base of pleon narrower than pereion,
giving interrupted outline (fig. 20B);
damp inland habitats only   *Ligidium*

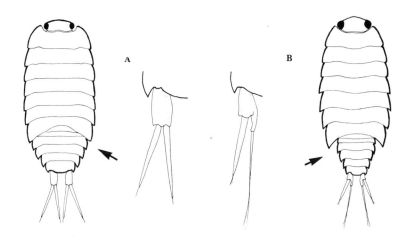

Fig. 20 Dorsal view of the pereion/pleon outline and uropod of the left side of
A *Ligia oceanica* and **B** *Ligidium hypnorum*.

GENUS *Ligia*
Ligia oceanica (Linnaeus) 1767; 25 mm; plate 2; figs. 2, 19, 20.
Our largest species; colour greyish-green, much mottled with black
in juveniles; eyes large and dark; head sunk into first segment of
pereion.
Native; crevices in rocks, caves, groynes, and quays; also in
sea-cliff grassland (up to 150 m on St Kilda); abundant in most
suitable localities right round coast, and in Ireland; strictly
nocturnal, feeding on shore detritus; *E13W4S4I2*.

86

GENUS *Ligidium*
Ligidium hypnorum (Cuvier) 1792; 10 mm; plate 3; figs. 19, 20
Dark mottled, with prominent head and eyes; very active; in the
field has strong superficial resemblance to *Philoscia muscorum* but is
easily distinguished by darker colour and numerous flagellar segments.
Native in south-east England; old records from as far north as Lancs
but nearly all recent records from East Anglia and Home Counties
(plate 8); confined to very damp habitats, characteristic of marshes
(abundant in the Fens), damp ditches, river banks, but also occurs in
forest litter on heavy soils; E9WoSoIo.

FAMILY Trichoniscidae
key to sub-families

1 Outline of pereion and pleon continuous
  (fig. 21A); dorsal surface with longit-
  udinal rows or ridges of tubercles   Haplophthalminae
  Outline of pereion and pleon not continuous
  (fig. 21B); spines and tubercles, if present,
  not arranged in rows   Trichoniscinae

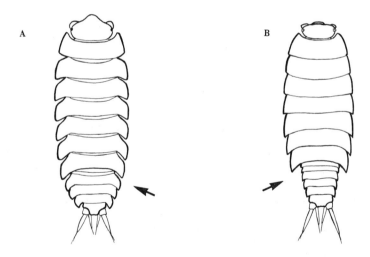

Fig. 21 Dorsal view of the pereion/pleon outline in **A** *Haplophthalmus danicus* and
B *Trichoniscus pusillus*.

SUB-FAMILY Haplophthalminae
key to GENUS *Haplophthalmus*

1 <u>Ridges</u> on dorsal surface of pereionites
   usually with 3–4 teeth (fig. 22A);
   any projections close to mid-line on
   3rd pleonite very feeble   *Haplophthalmus danicus*
   <u>Ridges</u> on dorsal surface of pereionites
   usually with more than 4 teeth (fig. 22C);
   prominent pair of projections on 3rd
   pleonite   *Haplophthalmus mengei*

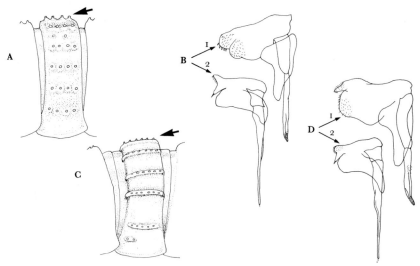

Fig. 22 Lateral view from the left side of the 2nd pereionite of **A** *H. danicus* and
C *H. mengei*.
First and 2nd male pleopods of the right side of **B** *H. danicus* and **D** *H. mengei*.

Haplophthalmus danicus Budde-Lund 1879; 4mm; figs. 21, 22
Dirty-white or yellow colour, eye difficult to see; body parallel-
sided and elongate; genitalia as in fig. 22B;
Native in Britain, doubtfully so in Ireland; rare; records since
1953 from Berks, Cambs, Worcs, and E. Yorks; past records much
more widespread; found in rotting wood and sometimes under garden
stones; *E4WoSoIo*.

Haplophthalmus mengei (Zaddach) 1884; 4 mm; figs. 19, 22
Same shape, size, and colour as *H. danicus*; genitalia as in fig. 22D.
Native in Britain, Ireland; rare, usually in soil; recent records
from Epping Forest, Surrey, Berks, Yorks, Durham; *E6WoSoIo*.

SUB-FAMILY Trichoniscinae
key to genera

1 <u>Eye</u> composed of 3 ocelli (fig. 23A,B), very
   close-set and fused in adult; body smooth and
   shining with sparse hairs (except rare alien
   Cordioniscus)   *Trichoniscus*
   <u>Eye</u> composed of 1 ocellus (fig. 23C), often
   obscure, sometimes absent; body neither
   smooth nor shining (except *Oritoniscus*)   *Androniscus*

Fig. 23 Dorsal view of the left side of the head to show the eye of **A** *T. pusillus*
(young juvenile); **B** *T. pusillus* (adult); and **C** *A. dentiger*.

GENUS *Trichoniscus*
key to species

1 <u>Colour</u> mottled purplish-brown (except very
   young animals)   *Trichoniscus pusillus*
   <u>Colour</u> white or yellow; few traces of
   pigment   *Trichoniscus pygmaeus*

Trichoniscus pusillus (Brandt) 1833; 3·5–5 mm; plate 4; figs. 19, 21, 23, 24
The only common reddish-brown species, mottled with white; brilliant
purple forms also occur; rear margin of telson feebly concave;
ocelli fused in adults, widely spaced in juveniles (fig. 23).
Native; our most abundant woodlouse but not the one most often seen
because of its small size and retiring habits; densities can be in

thousands per m² in thick grass litter; very prone to desiccation, migrates into the soil in drought and cold weather; occurs throughout British Isles up to 800 m in Scottish Highlands; occupies a great range of habitats, even in acid moorland; *T.p. pusillus*, the parthenogenetic triploid form, occurs in the north and west; *T.p. provisorius*, the bisexual form, occurs in the south-east (plate 8); the forms differ in size (triploid females reach 5 mm, diploid females 3·5 mm) and in genitalia (fig. 24A); $E43W6S14I14$. (*Oritoniscus flavus* may key out here because of its smooth body surface and its colour, but its eye, though large, has only 1 ocellus.)

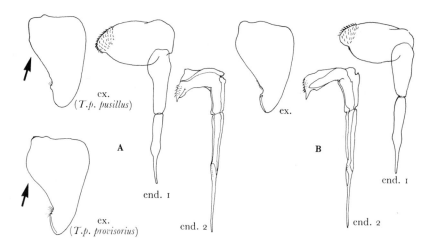

Fig. 24 Male genitalia of the right side of **A** *T. pusillus* and **B** *T. pygmaeus*, showing the exopodite of the 1st pleopod (ex.) and endopodites of the 1st and 2nd pleopods (end. 1 and end. 2).

Trichoniscus pygmaeus Sars 1899; 2·5 mm; fig. 24
Our smallest woodlouse; can be separated from juvenile *T. pusillus* (which also have very little pigment) by the latter's very large head and widely spaced ocelli (fig. 23A); rear margin of telson straight or feebly convex; both here and in *T. pusillus*, males are much smaller than females; genitalia as in fig. 24B.
Native; widespread and probably common but seldom reported because of its size and habits; soil species, usually found by turning over stones, but also occurs under deep litter in grassland, woodland, and gardens; $E10W0S1I1$.

Two rare aliens, both with a tuberculate body surface, key out here:

Cordioniscus stebbingi (Patience) 1907; 3 mm; genitalia as in fig. 25A;
hothouses in Glasgow and elsewhere; see Edney, 1953.
Cordioniscus spinosus (Patience) 1907; 5 mm; genitalia as in fig. 25B;
habitat and distribution as for *C. stebbingi*; see Edney, 1953.

GENUS *Androniscus*
Androniscus dentiger Verhoeff 1908; 6 mm; plate 5; figs. 23, 25
Colour in life ranges from white to deepest rose, fading rapidly in
alcohol; surface strongly tuberculate, each tubercle bearing a
central spine; genitalia as in fig. 25C.
Native in the south, probably only naturalized in the north, where
it is confined to gardens and waste ground; widely distributed in
Britain and Ireland and locally common; rare in Scotland; occupies
both very wet and very dry habitats, with a preference for limestone;
it is the most frequent species in caves and cellars; (*Androniscus
weberi* Verhoeff 1908, described in Edney's key, is now regarded as
a variety of *A. dentiger*); $E20W1S1I3$.

The following rare species and one alien key out here:

Oritoniscus flavus (Budde-Lund) 1906; 7 mm; colour and pattern as
for *T. pusillus* but appreciably larger; body smooth; eye, 1 large
ocellus; genitalia as in fig. 25D; native; a Lusitanian species;
occurs in Ireland only; old records mainly from the south-east;
few recent records; river banks, tumble-down walls, and moss;
$E0W0S0I3$.
Trichoniscoides albidus (Budde-Lund) 1879; 5 mm; colour in life
reddish-brown; genitalia as in fig. 25F; native; one recent record,
but widely recorded in the past; a soil species; $E1W0S0I0$.
Trichoniscoides sarsi (Patience) 1908; 3 mm; more or less white in
colour; genitalia as in fig. 25G; native; recent records only from
Sussex, Beds, and from caves in the Mendips; old records from
Scotland and from England except the south-west; mainly a soil
species; $E3W0S0I0$.
Trichoniscoides saeroeensis Lohmander 1923; 2·8 mm; little colour;
genitalia as in fig. 25E; native or naturalized, recently discovered
in two caves in Ireland, and in a mine at Warton Crag, Lancs (see
Sheppard, 1968); $E1W0S0I1$.
Miktoniscus linearis (Patience) 1908; 3 mm; genitalia as in fig. 25H;
an alien known only from Kew; see Edney, 1953.

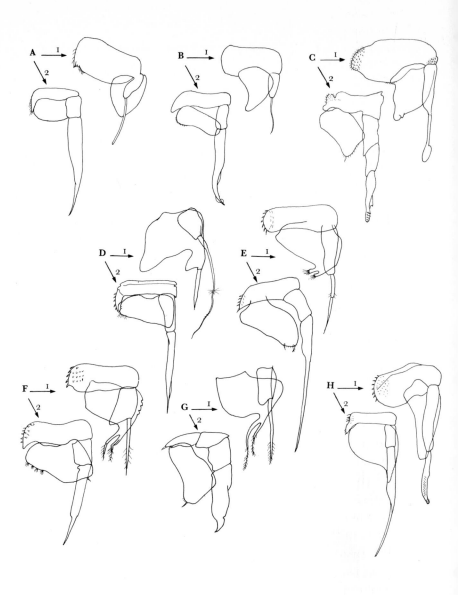

Fig. 25 First and 2nd male pleopods of the right side of **A** *C. stebbingi;* **B** *C. spinosus;* **C** *A. dentiger;* **D** *O. flavus;* **E** *T. saeroeensis;* **F** *T. albidus;* **G** *T. sarsi;* and **H** *M. linearis.*

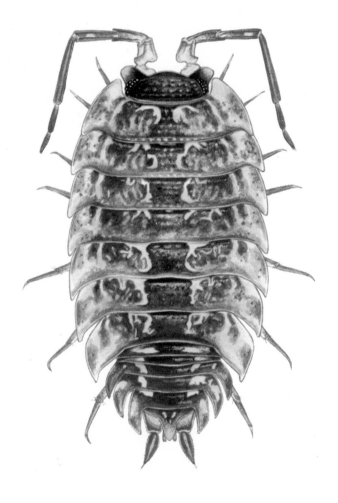

Plate 1 *Porcellio spinicornis* (female) from Aberford near Leeds, Yorkshire; length 10 mm.

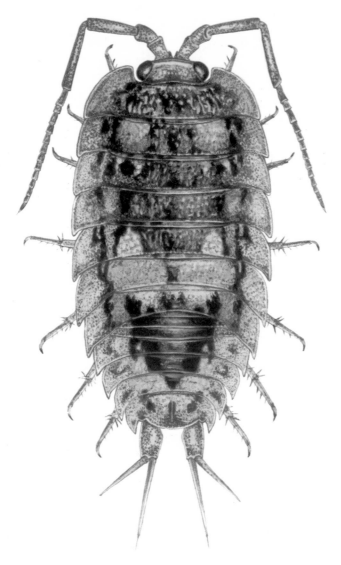

Plate 2 *Ligia oceanica* (female) from Robin Hood's Bay, Yorkshire; length 16 mm.

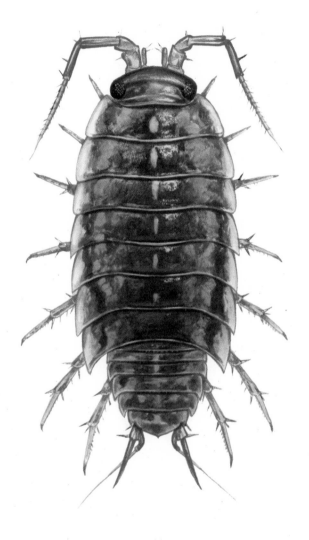

Plate 3 *Ligidium hypnorum* (male) from Godalming, Surrey; length 7 mm.

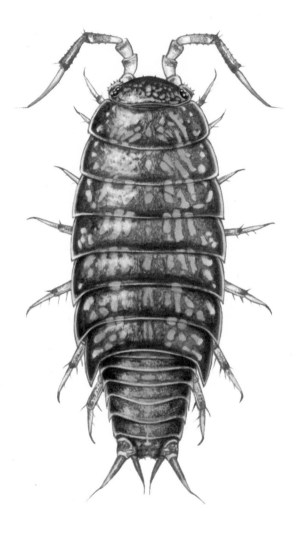

Plate 4   *Trichoniscus pusillus pusillus* (female) from Leeds, Yorkshire; length 3·5 mm.

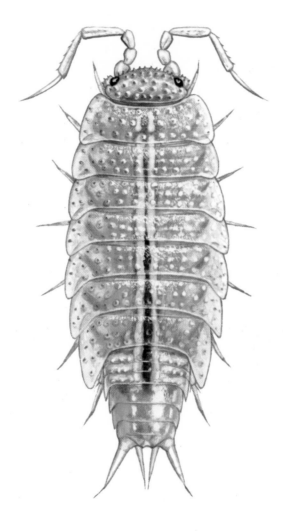

Plate 5 *Androniscus dentiger* (female) from Leeds, Yorkshire; length 4 mm.

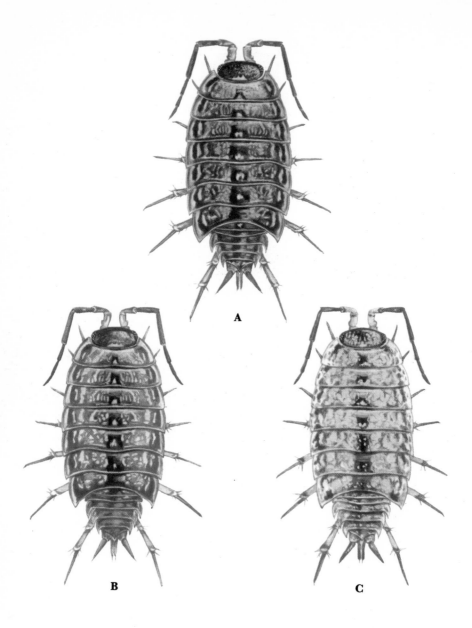

Plate 6 *Philoscia muscorum*: **A** typical form (female) from Silwood Park, Berkshire; **B** red form (male) from Silwood Park; **C** yellow form (female) from Hawsker, near Whitby, Yorkshire; lengths 6 to 8 mm.

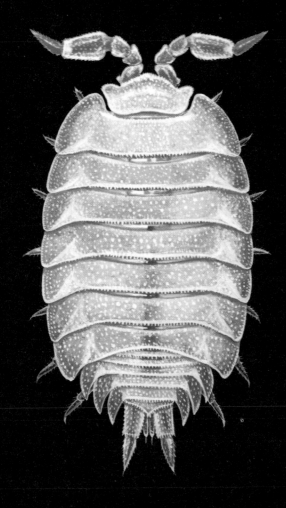

Plate 7 *Platyarthrus hoffmannseggi* (female) from nest of *Lasius flavus*, the Common
Yellow Ant, Box Hill, Surrey; length 2·5 mm. The foregut contains food
particles dyed red.

Metoponorthus cingendus

Ligidium hypnorum

Trichoniscus pusillus provisorius

Trichoniscus pusillus pusillus

Plate 8 Distribution maps. Dots represent vice-counties in which the species have been recorded since 1953. *M. cingendus* has a typically Lusitanian range; *L. hypnorum* a typically south-eastern one. *T.p. provisorius* (diploid) is mainly south-eastern, while *T.p. pusillus* (triploid) occurs everywhere except in the south-east

FAMILY Oniscidae
key to genera

1 Head with distinct lateral lobes (fig. 26A);
outline of pereion and pleon continuous; telson
ends in a long point (fig. 26E)   *Oniscus*
Head with feeble lateral lobes (fig. 26B,C);
outline of pereion and pleon not continuous;
telson with only slight point at tip (fig. 26D,F)   2

2 Exopodites of uropods slightly longer than
endopodites (fig. 26D)   *Philoscia*
Exopodites far longer than endopodites (fig. 26F)   *Halophiloscia*

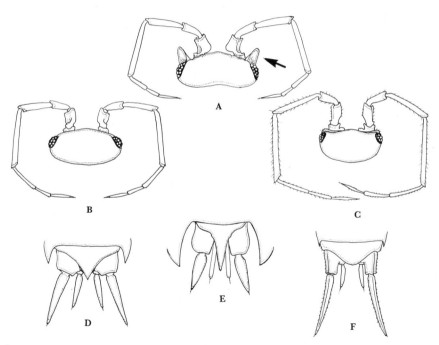

Fig. 26 Dorsal view of the head with antennae in **A** *O. asellus*; **B** *P. muscorum*; and
C *H. couchi.*
Dorsal view of the telson and uropods in **D** *P. muscorum*; **E** *O. asellus*; and
F *H. couchi.*

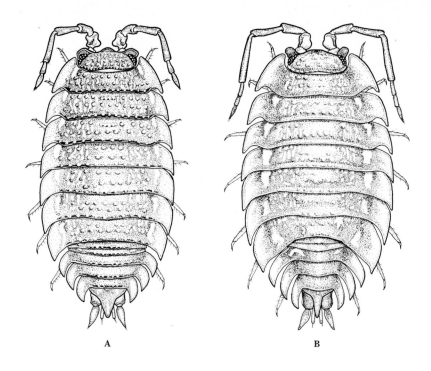

A   B

Fig. 27 Dorsal view of **A** *Porcellio scaber* (male), length 12 mm; **B** *Oniscus asellus* (male), length 13 mm.

GENUS *Oniscus*
Oniscus asellus Linnaeus 1761; 16 mm; figs. 6, 26, 27
Usually grey with irregular lighter patches, but both yellow and orange forms common near the sea; body surface bears irregular raised welts; body glossy in adults but often rough in young animals; latter have light or orange patch on epimera of last pereionite.
Native; our most commonly seen species; ubiquitous, found all over British Isles; favourite habitat rotting wood; abundant in gardens, waste ground, and coastal grassland; avoids drier habitats such as limestone turf; one of the few woodlice that can tolerate acid soils; $E45W7S14I21$.

GENUS *Philoscia*
Philoscia muscorum (Scopoli) 1763; 11 mm; plate 6; figs. 19, 26
Very active; usually brown with darker head and median stripe
along body; red forms and yellow forms occur; confusion with
Ligidium hypnorum and *Metoponorthus cingendus* possible.
Native; one of our commonest species; occurs all over British
Isles but sparse in north of England and in Scotland; character-
istic of hedgerows and grassland but does penetrate woodland
and gardens; tolerates surprisingly dry conditions and is
abundant in dune grassland; $E45W4S6I19$.

Two rare hothouse aliens key out here:

Chaetophiloscia meeusei Holthuis 1947; 8 mm (not 1 mm as stated
by Holthuis); from *Philoscia* by mottled purplish colour, slender
antennae and narrower pleon, giving a highly discontinuous
pereion/pleon outline; only from Kew; see Holthuis, 1947.
Chaetophiloscia patiencei (Bagnall) 1908; 3 mm; distinguished by
minute size; Kew; Winlanton, Durham; see Edney, 1953.

GENUS *Halophiloscia*
Halophiloscia couchi (Kinahan) 1857; 10 mm; fig. 26
Colour like *T. pusillus*; very long antennae.
Native, but rare, littoral species; past records from Cornwall
to Sussex, but recently reported only in the Hartland area of
North Devon, where it occurs under stones around the high water
(springs) level on a rocky shore; $E1WoSoIo$.

FAMILY Squamiferidae
GENUS *Platyarthrus*
Platyarthrus hoffmannseggi Brandt 1833; 3·6 mm; plate 7
Small, oval, white, and blind; stubby antennae with swollen last
peduncular segment.
Native, living in ants' nests, chiefly those of *Lasius flavus*
(the common yellow ant); can live away from ants' nests but
seldom found on its own in the field; mainly in the south-east,
north to Yorks; perhaps restricted to calcareous soils; $E22W2SoI1$.

A rare alien keys out here:

Trichorhina tomentosa (Budde-Lunde) 1893; 5 mm; nearly always has
eyes; antennae not swollen; Belfast; see Gruner, 1966.

FAMILY Porcellionidae
key to genera

1 Pseudotracheae present on all five pairs of pleopods
   (*not* easily visible in alcohol) (fig. 28B)   *Trachelipus*
   Pseudotracheae present only on first two pairs
   of pleopods (*not* easily visible in alcohol)
   (fig. 28A)   2

2 Side-lobes of head strongly developed (fig. 28D);
   outline of pereion and pleon continuous   *Porcellio*
   Side-lobes of head poorly developed (fig. 28C);
   outline of pereion and pleon weakly discontinuous   *Metoponorthus*

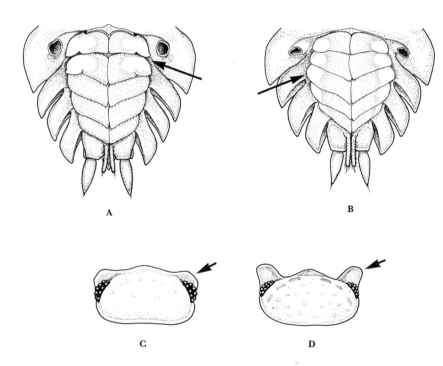

A   B

C   D

Fig. 28 Ventral view of the pleon (female) to show the distribution of pseudo-
        tracheae (light areas) in **A** *P. scaber* and **B** *T. rathkei*.
        Dorsal view of the head in **C** *M. pruinosus* and **D** *P. dilatatus*.

96

GENUS *Trachelipus*

Trachelipus rathkei (Brandt) 1833; 15 mm; figs. 19, 28, 29.
Mottled yellowish-grey with 3 distinct double stripes along pereion;
head with angle between central and lateral lobes always shallow
as in fig. 29C; exopodite of second male pleopod as in fig. 29A.
Native; recent records suggest south-east distribution in Home
Counties, Hunts, Berks; old records in England and Scotland suggest
wider distribution; juveniles probably of this species recorded
from St Kilda in 1961; *E4WoS1Io*.

Another species of *Trachelipus* and two rare aliens key out here:

Trachelipus ratzeburgi (Brandt) 1833; 16 mm; distinguished from
T. rathkei by acute angle between side and central lobes of head
(fig. 29D); exopodite of second male pleopod as in fig. 29B; no
confirmed recent British records; old specimens given to British
Museum (Nat. Hist.) are in fact *T. rathkei*; *EoWoSoIo*.
Nagurus cristatus (Dollfus) 1881; 10 mm; head as in fig. 29E;
distinguished from following species by size; Northumberland only;
see Schmölzer, 1965.
Nagurus nanus Budde-Lund 1908; 4·5 mm; hothouse in Belfast
Botanic Garden; one specimen only; see Foster, 1911.

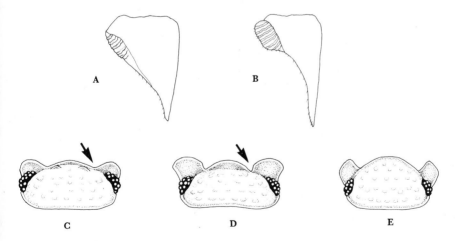

Fig. 29 Exopodite of the 2nd male pleopod of the right side in **A** *T. rathkei* and
B *T. ratzeburgi*.
Dorsal view of the head in **C** *T. rathkei*; **D** *T. ratzeburgi*; and **E** *N. cristatus*.

GENUS *Porcellio*
key to species

1 <u>Bo</u>dy smooth, not tuberculate; uropods long
  (fig. 30A)   *Porcellio laevis*
  <u>Bo</u>dy rough, covered with tubercles; uropods
  short (fig. 30B,C,D,E)   2

2 <u>Telson</u> appears elongate, usually with a
  round tip (fig. 30D,E)   *Porcellio dilatatus*
  <u>Telson</u> short with pointed tip (fig. 30B,C)   3

3 <u>Head</u> colour dark, contrasting with lighter
  pereion; the latter blotched with a broken
  but distinct median stripe; heavily
  tuberculate on head only; animal never a
  uniform dark slate colour (plate 1)   *Porcellio spinicornis*
  <u>Head</u> colour not darker than pereion which,
  if blotched, never has median stripe
  (fig. 27A); adults often a uniform dark
  slate colour; very heavily tuberculate on
  pereion as well as on head   *Porcellio scaber*

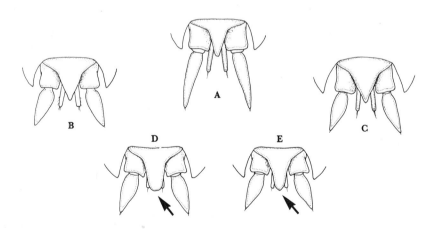

Fig. 30 Dorsal view of uropods and telson of **A** *P. laevis*; **B** *P. spinicornis*; **C** *P. scaber*; **D** *P. dilatatus* (normal); **E** *P. dilatatus* (variant).

Porcellio laevis Latreille 1804; 18 mm; fig. 30
The only brown, uniformly smooth, glossy *Porcellio*.
If native, only in a few localities; mostly naturalized, closely
associated with man; widespread but rare; well established in the
compost heap of the Botanic Garden in Oxford; other records
from Somerset, Kent, London (Shepherds Bush); old records
from Scotland and Ireland; *E6WoSoIo*.

Porcellio dilatatus Brandt 1833; 15 mm; figs. 28, 30
Grey-brown, with well-defined broad stripes on each side; char-
acteristic telson shape obvious in large adults, much less so in
young animals which, except for pattern, could be confused with
P. spinicornis or *P. scaber*.
Native in southern England, probably naturalized in the north;
common in chalk grassland in Beds; other recent records from
Monmouth and Yorks; *E4W1SoIo*.

Porcellio spinicornis Say 1818; 12 mm; plate 1; fig. 30
Median lobe of head usually forms a smooth curve, but may be almost
pointed, or trapezoid.
Native in England, perhaps only naturalized in Scotland; native
habitat distinctive, occurs on limestone walls, quarry faces
and cliffs in open sites; also on old ruins and in buildings where
mortar or limestone has been used, even in city centres; recent
records mainly from southern England, but north to Durham, IOM,
Glasgow and Fife; *E13WoS2Io*.

Porcellio scaber Latreille 1804; 17 mm; figs. 19, 27, 28, 30
Median lobe of head comes to a blunt point; base of antennae commonly
orange; yellow, orange forms, speckled with black, are frequent,
particularly in juveniles and females.
Native; as ubiquitous and common as *Oniscus asellus* but likes
drier habitats, including sand dunes and acid heaths; common on tree
trunks and walls; abundant on waste ground, in gardens, and in
maritime grassland; *E50W7S14I18*.

A rare alien keys out here:

Agabiformius lentus (Budde-Lund) 1885; 6 mm; dorsal surface much
more strongly curved than in other Porcellionidae; very short antennae;
background colour apparently whitish, with 3 broad and ill-defined
stripes; the least rare of the aliens; hothouses only; see Vandel, 1962.

GENUS *Metoponorthus*
key to species

1 Colour purplish-brown with a blue-grey
   'bloom'   *Metoponorthus pruinosus*
   Colour dark mottled   *Metoponorthus cingendus*

Metoponorthus pruinosus (Brandt) 1833; 12 mm; fig. 28
Colour unique, bloom easily rubbed off and disappears in alcohol.
Native or naturalized; characteristic of farmyards, particularly
manure heaps; a few widely scattered recent records in England;
much more widely recorded in the past; *E*10*W*0*S*0*I*0.

Metoponorthus cingendus (Kinahan) 1857; 7 mm
In the field resembles *P. muscorum* with which it occurs, but is
darker, slimmer, and has no median stripe; each of the anterior
segments of the pereion has a transverse *carina* (ridge).
Native; a Lusitanian species (plate 8), found in south-west
England, South Wales, and in Ireland, where it is prevalent in
the coastal counties of Eire, and seems to replace *P. muscorum*
in some sites. In Britain, never found more than 100 m from the
shore, but penetrates well inland in Ireland; *E*4*W*1*S*0*I*7.

A rare species keys out here:

Acaeroplastes melanurus (Budde-Lund) 1885; 6 mm; colour dark-
mottled with a very evident central stripe; pereion segments without
carinae; native; known only from the Hill of Howth near Dublin; has
not been seen for 60 years; Lusitanian and Mediterranean
distribution abroad; *E*0*W*0*S*0*I*0.

FAMILY Cylisticidae
GENUS *Cylisticus*
Cylisticus convexus (De Geer) 1778; 15 mm; figs. 19, 31
Broad grey stripes; tubular uropods prominent in life (fig. 19H);
can roll up but seldom does so (fig. 31C).
Native in the south, naturalized in the north; rather rare; likes
dry situations; native habitat chalk downs, also known from sea-shore
drift line; now closely associated with man in places, it is well
established on waste ground in the centre of Leeds; only two recent
records from Scotland (Wigtown and Fife); old Irish records few but
widespread, no recent records; *E*7*W*0*S*2*I*0.

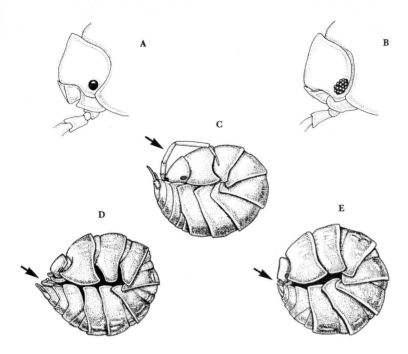

Fig. 31 Lateral view of the left side of the head showing the eye in **A** *E. purpurascens* and **B** *A. vulgare.*
Lateral view of the rolled up attitude in **C** *C. convexus*; **D** *A. depressum* and **E** *A. vulgare*; (all approximately 5 mm in diameter).

FAMILY Armadillidiidae
key to genera

1 Eyes of a single ocellus (fig. 31A); County
  Dublin only   *Eluma*
  Eyes compound (fig. 31B)   *Armadillidium*

GENUS *Eluma*
Eluma purpurascens Budde-Lund 1895; 9 mm; fig. 31
Colour mottled purplish-brown; very long endopodites of uropods
visible from above.
Native; a very rare Lusitanian species found only in drift-line
debris on a few beaches in County Dublin; *EoWoSoI*1.

GENUS *Armadillidium*
key to species

1 <u>Head</u> viewed from front as in fig. 32B   *Armadillidium nasatum*
  <u>Head</u> viewed from front as in fig. 32C   *Armadillidium depressum*
  <u>Head</u> viewed from front as in fig. 32D   *Armadillidium vulgare*

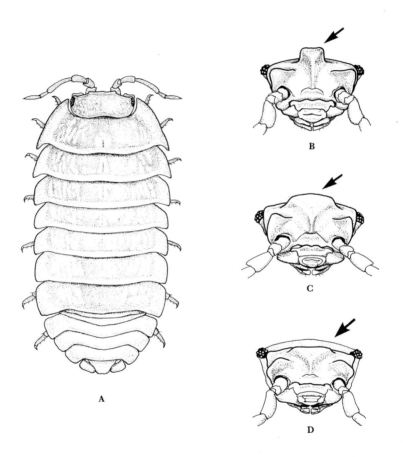

Fig. 32 **A** Dorsal view of *A. vulgare* (female), length 11 mm.
      View of head from the front in **B** *A. nasatum*; **C** *A. depressum*; and **D**
      A. vulgare.

Armadillidium nasatum Budde-Lund 1885; 21 mm; figs. 19, 32
Scutellum (snout) extremely well developed, seen from front tall but
narrow, less pronounced in young juveniles but still evident.
Native in the south on limestone; likes hot sunny slopes and screes;
recent records from Cheddar Gorge, Surrey, Mon, Glos; occurs in
greenhouses in the north; probably native in Ireland; *E6WoSoIo*.

Armadillidium depressum Brandt 1833; 20 mm; figs. 31, 32
Scutellum very wide, projecting above top of head; colour slate-grey.
Native; very rare except in south-west, where it is known in a few
localities in Devon, Som, Dorset, and Mon; often found there in
gardens; *E5WoSoIo*.

Armadillidium vulgare (Latreille) 1804; 18 mm; figs. 2, 19, 31, 32, 33
Scutellum narrower than in *A. depressum*, not projecting above top of
head; colour often dark slate-grey but very variable; red and varie-
gated forms known; genitalia and 7th male pereiopod as in fig. 33A,E.
Native; the only widespread and common pillbug; restricted to calcareous soils except in coastal habitats; very common in south-east,
local in western and northern England, rare in Scotland, absent from
north-west Ireland; preferred habitat stony turf on chalk or limestone;
often seen in the sun; *E39W3S1I11*.

The following rare species and one alien key out here:

Armadillidium pictum Brandt 1833; 7 mm; genitalia and 7th male pereiopod as in fig. 33B,F; native, recent certain records only from the
southern Lake District, where it has been found on limestone
pavement: *E2WoSoIo*.
Armadillidium pulchellum (Zencker) 1799; 3·5 mm; distinguished by
small size, genitalia and fringe of bristles on 7th male pereiopod
(fig. 33C,G); native; very rare; rediscovered in 1971 in southern
Lake District, among Juniper on limestone; unconfirmed recent
record from moorland in Snowdonia; old records from other
localities in England and Ireland; *E2WoSoIo*.
Armadillidium album Dollfus 1887; 6 mm; distinguished by genitalia
and projection on the 7th male pereiopod (fig. 33D,H), and by
sprinkling of small spines on surface (obvious on dry specimens);
native; only one old record, but now known from 14 sites on the west
coast of England from Devon to Cumberland, from Norfolk and
Yorks, and from 3 counties in south-east Ireland (Harding, 1968);
in old drift-line debris on sandy shores; *E4W3SoI3*.

Reductoniscus costulatus Kesselyák 1930; 2·2 mm; the only alien pill-bug, and the only one with 5 (rather than 2) pairs of pseudotracheae; from the Victoria House at Kew; see Holthuis, 1947.

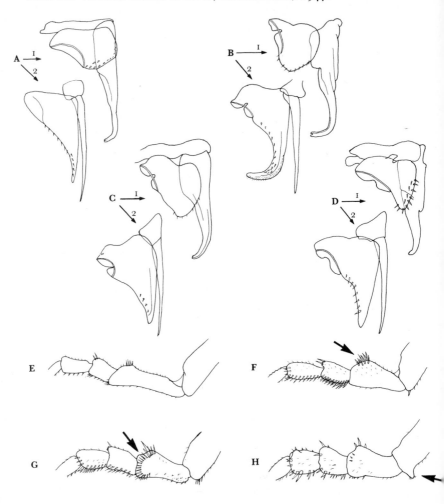

Fig. 33 First and 2nd male pleopods of the right side in **A** *A. vulgare*; **B** *A. pictum*; **C** *A. pulchellum*; and **D** *A. album*.

Anterolateral aspect of the 7th male pereiopod of the right side in **E** *A. vulgare*; **F** *A. pictum*; **G** *A. pulchellum*; and **H** *A. album*.

9 TECHNIQUES OF STUDY

In this Chapter I want to concentrate on practical matters, giving some idea of how woodlice can be collected and extracted, how they can be cultured, and how books and papers about them can be obtained.

Collecting

The simplest form of collecting is by hand-searching, which means getting down on hand and knees, peering into the litter, and sucking up the woodlice with a pooter (otherwise known as a bug-sucker or aspirator; fig. 34). Useful aids in this sort of activity are a widger, a bread knife, and a large white Polythene sheet or tray. The widger is a gardening tool used for transplanting seedlings and is invaluable for

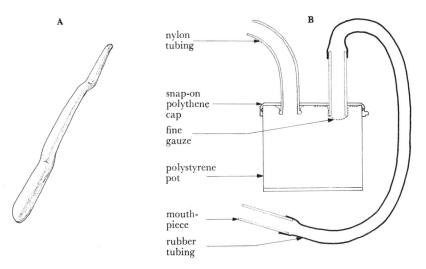

A

B

nylon tubing

snap-on polythene cap

fine gauze

polystyrene pot

mouth-piece

rubber tubing

Fig. 34 Tools of the trade: **A** the widger and **B** the pooter (basically a polystyrene specimen jar with a snap-on cap).

poking into vole tunnels or prising up bark. The bread knife is for cutting through grass tussocks at root level so that the whole tuft can be shaken over the white sheet, whereupon the occupants fall out and can be pooted up. This is much more effective than trying to spot the animals in the standing tussock. Note that large woodlice tend to get stuck in the pooter intake tube and are best picked up with a pair of forceps; remember though that they are rather brittle and require careful handling. Collecting *Ligia* is best done by hunting them down in the open after dark, but they can be eased from their crevices during the day with a piece of bent wire.

In hand-sorting in the field one tends to miss the smaller animals and it is much better to collect up the litter or whatever and do the sorting indoors with the aid of a bright lamp. Besides giving good lighting this heats the animals up and makes them restless so that they move about and are more easily seen. The material should be sorted out, a little at a time, in a large white tray.

For the smaller woodlice, particular the deeper living ones, hand-sorting should be avoided because so many animals will be missed; instead, samples should be put through an extraction apparatus as described in the next section. Extractors should also be used where data are needed on density, biomass, or size structure, as when using methods described in the section on quantitative sampling. Where you want only a steady supply of animals for experiments or want to study rates of dispersal using marked animals, 'cryptozoa boards' are useful. These are flat boards of seasoned but untreated timber which are left out in the study area and under which woodlice gradually come to shelter. On waste ground bricks make effective and unobtrusive 'boards'.

A technique of collecting that can be recommended only with strong reservations is pitfall trapping. Pitfall traps, in the form of jam jars or cans buried up to the lip in the ground so that animals running about on the surface fall into them are easy to use but give very misleading information about numbers. This is because the catch of any species will depend on abundance, activity, and agility, and it is impossible to estimate one without some independent measure of the other two. The accompanying table shows what can happen. The

	number in pitfalls (May, 1964)	live weight of large adults (mg)	density per m^2 (estimates from soil/litter cores)
Trichoniscus pusillus	19	1·5	831
Philoscia muscorum	22	18·0	75
Armadillidium vulgare	40	100·0	12

pitfall results, if taken as an indication of abundance, would be spectacularly wrong, as is indicated by the true density figures which were found by core sampling. Nor, if one knows this density, can one conclude that the difference is caused by activity because, in fact, a great deal of the difference is caused by differences in agility between species. *A. vulgare* is a large, heavy, bumbling animal that easily falls into pitfalls whereas *T. pusillus*, being very much smaller and lighter, is far less likely to tumble in because it can stop itself on the edge. Pitfall traps may be used for comparing the relative activity of one species at different times of the year or, at a pinch, for preliminary surveys of larger woodlice, but otherwise they should be avoided.

For experiments involving living animals, *Porcellio scaber* and *Oniscus asellus* are the easiest animals to use because they are quite large and very common all over the country. *P. scaber* often abounds in the drier parts of old compost heaps, under piles of rubbish, or under the loose bark of dead trees. *Oniscus asellus* is equally common but in damper spots, particularly under stones or the bark of well-rotted logs. Both species flourish, along with *Androniscus dentiger* and *Trichoniscus pusillus*, under the rubble and debris of slum clearance sites, if these have had a few years to mature. Overgrown graveyards are also very good collecting areas from some points of view, at least. A very important point to remember is that, in winter frosts, these species go well down into the ground and are best looked for by turning over really large logs and stones, down to a metre in depth.

In the north of England some useful species, such as *A. vulgare*, are rare in the open, but can often be found in hot-houses.

Extracting

Devices for extracting animals from soil or litter may be classed as *dynamic* or *passive*. Dynamic methods, such as the tried and trusty Tullgren funnel, create gradients of heat and humidity which cause the animals to evacuate the soil or litter and to drop into the collecting fluid. In its crudest form such an apparatus consists of a steep-sided funnel with a collecting tube of alcohol at the narrow mouth below; the sample is supported just below the wide mouth at the top by a mesh (see fig. 35). An overhead light provides the drying effect. Simple funnels for leaf litter can be constructed from plastic or aluminium sheeting, and large plastic funnels can be bought. The sides should be steep so that the animals do not collect on them—an angle of about 60° is needed. Note that if the light is too intense the sample will dry too quickly and trap the animals or may even catch fire, bringing the experiment to an untimely end.

Refinements of this basic apparatus are legion and mainly involve control of humidity conditions in the funnel, increasing the humidity gradient by cooling the funnel bottom, and replicating the units so that statistical analyses can be applied to the results. Reviews of the most advanced dynamic methods such as the Macfayden air-conditioned funnel, and the Kempson-Lloyd-Ghelardi infra-red apparatus will be found in Southwood (1966) and Wallwork (1970). The latter apparatus can be built on a do-it-yourself basis and is very suitable for woodlice but it does need a cooling system of running water.

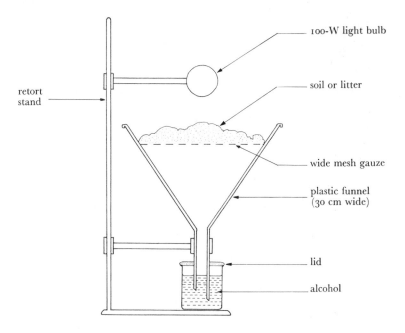

Fig. 35 A simple Tullgren funnel for extracting animals from soil and litter. The mesh should be coarse enough to let the larger woodlice through.

Passive or mechanical methods of extraction are all those which do not require the animals to extract themselves. Under this heading come serving, flotation, and the grease-film apparatus. There is little point in giving any details here as most are unsatisfactory for woodlice, although perhaps this is because little attention has ever been given to their use in this direction.

Quantitative sampling

Where population data are required, collecting must be done in an organized way, preferably using extraction apparatus and samples of standard area and depth. Where the terrain makes this impossible (as, for example, when sampling *Ligia* in shingle) standard time counts may be used, but you have to guard against bias towards the larger animals which are easier to see and to catch and your results will be difficult to compare with those of other people. Usually, it is possible to take soil-cores or quadrats of known area. In population studies, each sample should consist of a number of sampling units or replicates. The total area sampled during the study should be small in relation to the total size of the study area so that the removal of material does not affect the remaining population. The number of replicates determines the accuracy with which density is estimated—the more replicates, the greater the accuracy. Where individuals are as highly aggregated as woodlice, there will be a great deal of variation between replicates and hence wide confidence limits on density estimates. With woodlice populations, using between 10 and 20 replicates, one usually has to settle for indications of a trend in numbers over the year rather than statistically significant differences from month to month, except when the young are set free.

In size, each replicate or sampling unit should be large enough so that all, or nearly all, of them contain some animals. The upper limit on individual size is set by the capacity of the extraction equipment and by the need to have enough replicates to treat the results statistically. The largest feasible size of core, using a thin walled corer, is $1/50 \text{ m}^2$ (16·0 cm diameter). Anything larger in grassland can be cut out as a square using steel plates hammered in along the edges to cut the binding roots and contain the animals. A similar device can be used as a sampling unit in forest litter to isolate the material while it is put into bags (soil and litter preferably being treated separately). Replicates smaller than $1/50 \text{ m}^2$ will be too small for most woodlouse populations.

The total size of each sample will be a compromise between the need for large numbers of animals (making it large) and feasibility (making it small). In practice, to study size structure one might need between 50 and 200 individual animals to recognize a pattern, depending on the number of size classes chosen and how sharply the generations are divided in size; the total size of the sample should be adjusted to achieve this target.

For a long-term study, regular samples of a constant number of replicates should be taken from a sampling area marked out in compartments, several replicates being taken at random, using a table of random numbers, from each compartment. A *stratified random* sampling arrangement of this kind prevents personal bias on the choice of sampling sites and ensures that not all replicates come from one corner, as can happen if samples are taken at random from the study area as a whole. 'Clustering' of this kind can be very damaging because sites are never homogeneous and one corner may be very different from the rest of the area. Details of statistical methods for applying to sampling data can be found in Dowdeswell (1959), Bailey (1959), and Snedecor and Cochran (1967).

Preserving

Woodlice are best preserved in 70 per cent alcohol. The addition of glycerol is usually recommended, but I find it makes the animals very brittle. Preserved woodlice should be stored in small tubes plugged with cotton-wool, with a label inside written in Indian ink. This should give details of date of capture, locality, habitat, and the name of both collecter and identifier. A number of small tubes can be stored *up-side-down* in a large, wide-mouthed jar full of 70 per cent alcohol. Specimens which dry out can be resurrected by careful treatment with KOH (less than 10 per cent concentration) or (preferably) with Na_3PO_4 (near saturated solution for 48 hr minimum); then distilled water (to wash); then alcohol.

Measuring and sexing

So far as dead specimens are concerned, using length as a measure is very bad policy with woodlice because of the amount of expansion or contraction between segments which takes place at death; it can amount to as much as 50 per cent of the living length. A better approach is to measure head width because the head capsule is a single skeletal element which cannot change shape. A microscope with a graduated eyepiece has to be used, but this is the only way to ensure accuracy. Taking measurements of live animals is quite easy in the case of *Philoscia* and *Trichoniscus* which, provided the light is not too strong, move about very little if put in a Petri dish with damp filter paper. Pillbugs are impossible, however, because they will not keep still. Narcotizing them with CO_2 may bring some success but they recover very quickly and start moving about again.

Weighing woodlice, as a prelude to biomass estimation, is a very finicky business because they can weigh less than 0·0001 g. Also, if alive, they lose water so quickly that their weight changes rapidly. Although this difficulty can be avoided by using a miniature stoppered bottle, it is more convenient to use dry-weight rather than live-weight, which also allows biomass to be compared more meaningfully with that of other animals.

For dry-weights, animals should be kept in an oven at 40° C (anything higher may drive off volatile fats) until a constant weight is reached indicating that all the water has been driven off. Unless a very fine balance is used (accurate to 0·00001 g) small woodlice have to be measured in batches and the average weight taken. Rather than weighing animals from every sample, it is convenient to find the relationship between size and weight from one sample (expressed graphically or as a regression) and to use this in calculating biomass subsequently. Note that such biomass figures include the weight of the exoskeleton as well as the weight of tissue. To find the latter (for energy flow studies) the animals must be burnt in a muffle furnace to allow the ash-free dry-weight to be calculated. This is the most useful measure of biomass.

Sexing woodlice is a matter of looking for male genitalia or female brood pouches. Non-breeding adult females can be distinguished from juvenile males only by size, but fortunately the rudimentary male genitalia become visible quite early in development so that size is a reliable guide. Maturity in males is best judged by looking for spermatozoa. They appear as fine glistening strands when the seminal vesicles are opened. Some woodlice can be identified only if the genitalia are dissected out; this is best done with the aid of fine mounted needles and a binocular microscope. Valuable characters are found on the exopodites and endopodites of the first 2 pairs of pleopods (fig. 6). When teased from the body the genitalia are transferred with a pipette to a microscope slide and mounted in glycerol (for a temporary preparation) or in DPX or, by way of absolute alcohol and xylol, in Canada balsam for a permanent preparation. Mandibles, which require removal to count the number of penicils (fig. 3) are treated in the same way, but it is absolutely vital to know which is the left mandible and which the right (bearing in mind that the animal will be dissected from underneath). Confusion here can lead you to the wrong genus, using some keys.

Examination of chromosomes is another delicate operation but, in fact, a very easy one. Chromosome counts have to be made when studying the two forms of *Trichoniscus pusillus*, in which species the number of chromosomes is low enough ($2n = 16$) to make counting during the metaphase of mitosis fairly easy. The secret is to use very young embryos, where cell division has just begun and the chromosomes are very large. Such embryos in the brood pouch have a clear yellow colour and show no sign of differentiation. Each embryo should be teased out on a microscope slide from the brood pouch, its membranes split, and the contents stained in a drop of lacto-orcein (1 per cent orcein in 4·5 per cent lactic acid) for a minute or two. After cleaning the slide of debris, the preparation should be squashed firmly with a coverslip and the edges sealed with a rubber solution such as Cow Gum to make a preparation which will last some days.

Fixing, sectioning, and staining
Woodlice are not very suitable for histological work because of difficulties in fixation and because of the hardness of the cuticle which causes tearing during sectioning. This latter problem can be avoided to some extent by using species with a soft cuticle such as *Trichoniscus* or young specimens of the more heavily armoured species such as *Porcellio*.

A satisfactory fixative is alcoholic Bouin (150 cm^3 of 80 per cent alcohol; 60 cm^3 of 40 per cent formalin; 15 cm^3 of glacial acetic acid; 1 g of picric acid). Fixation is aided by cutting off the head or tail of the animal as appropriate to allow the fixative to penetrate.

For sectioning, a hard (58° C) paraffin wax-plastic works well. Cut at 10 μ for general work but preferably at 5 μ for cellular details. Haematoxylin and eosin staining is very satisfactory but for really good differentiation Mallory's triple stain can be used. The recommended procedure is as follows: sections to water; $\frac{1}{2}$ minute in 0·25 per cent aqueous solution of acid fuchsin; rinse without washing and transfer for 2 hours to solution of anilin (methyl) blue 0·5 g, Orange G 2 g, phosphotungstic acid 1 g, all in 100 cm^3 water. The times quoted should be altered to suit the conditions.

Culturing
The most important thing about keeping woodlice is that conditions should not become too wet or too dry. Condensation is a sign that conditions are too wet, except for the most hygrophilic species like *Ligia, Ligidium,* and *Trichoniscus.* At the other extreme the pillbugs and some *Porcellio* species require quite dry conditions, as is indicated by the dryness of their natural habitats. Humidity can be maintained by putting a lump of wet cotton-wool in one corner of the culture to create a humidity gradient across the floor, so that the animals can choose their optimum humidity. Even better, the bottom of the container can be filled with a thick layer of plaster-of-paris which, being porous, soaks up a lot of water and maintains a steady humidity for a long time. Another good culture medium is fine-grained Vermiculite kept moist and covered with tiles.

Woodlice can tolerate a wide range of temperature but violent changes can cause mortality, as can exposure to direct sunlight, so that cultures should always be kept in the shade at fairly stable temperatures. In any case, if breeding and growth-rates are being studied, a known constant temperature is necessary if you wish to compare your results with other people's because these rates vary with temperature. Even better than a fixed temperature, because it is nearer natural conditions, is a regime with 24 hour fluctuations between a constant maximum and minimum. For food, slices of carrot and potato make up the time-honoured recipe and seem quite suitable in the short term, but such an unnatural diet may explain why so many cultures die off after several weeks or months for no apparent reason. Far better to provide some of the litter in which the animals were found because that is more likely to provide them with their dietary requirements. Whatever the food-stuff, a combination of high humidity and high temperature can cause a rapid growth of mould which in some way—perhaps simply by entanglement—kills the animals. While thinking of food it is worth mentioning one implication of Wieser's work on copper metabolism, which is that woodlice put into a culture without a supply of their own faeces may not be able to obtain sufficient copper to maintain their reserves. If the loss is severe or sufficiently prolonged presumably they will die. Other additives which may be beneficial are chalk and filter paper. The latter is certainly eaten with great relish, although whether it has any effect is not known. A liking for paper extends to any data labels put into a culture—they don't last long!

Cultures can be kept in anything from a small stoppered tube to a large plastic dustbin, although where the container is open it should be smooth sided so that the captives cannot climb out. A large culture can be kept going very well in a bin half-full of litter, if it is watered occasionally. It is best to have some drain holes in the bottom as an insurance against over-enthusiastic watering. For culturing in the field, as when measuring individual growth-rates, one can use small yoghurt pots with fine mesh nylon over the top and bottom. These can be left buried in the ground for many months, but if small mammals become inquisitive the nylon mesh will have to be replaced with phosphor-bronze or perforated zinc.

A problem in crowded cultures is cannibalism of soft and fairly immobile individuals during the moult, but this can be reduced by pro-viding more shelter in the form of pieces of bark, stone, or leaf litter.

Marking animals

Because the cuticle is porous and never dry, woodlice are exceedingly difficult to mark with paint, which tends to flake off. This applies particularly to smooth-surfaced species. Quick-drying paints with an acetone base (aeroplane dope, for example) are quite successful on rough-surfaced species such as *P. scaber* although, like all surface marks, they are lost at the moult. Water-soluble paints, such as 'Humbrol' colours for toy soldiers, are absorbed by the cuticle and are more effective. 'Magic-marker' pens seem to work well except that the colour may become very faint after a few days.

Accessibility of literature
For people who do not have access to a university library, obtaining the specialist books and papers cited in this text can be quite a problem. There is, however, one effective answer in the United Kingdom and this is to make use of the National Lending Library for Science and Technology. This has an extremely large stock of books and periodicals which it sends by mail, and it would certainly be able to provide all the references quoted in this book except for unpublished theses; these will usually be made available by the parent universities on request. Note that the National Lending Library lends only to other libraries and application must be made through a public or approved library, who are responsible for the book until it is returned. Where a paper is needed for frequent reference it is legally possible, in most cases, to take a photocopy; (alternatively a reprint can be obtained by writing to the author).

Papers on woodlice come out in such a wide variety of journals that the only way to keep track of them is through the abstracting services such as *Biological Abstracts*. This is published monthly and consists of titles and short summaries of recently published papers. It can, with practice, be scanned quite rapidly, but owing to its size and price it is available only in specialist libraries. The same is true of the Zoological Record, which, although much less up to date, is easier to use.

Finally, for general assistance and advice, it is well worth consulting the staff of the British Museum (Natural History) who are both willing and well qualified to help.

INVESTIGATIONS

The suggestions for practical work given here are of two types: 1, experiments worked out in detail, designed to involve a whole class and using readily available apparatus; 2, suggestions for further work. The latter are, in most cases, more suitable for individuals or small groups. The criteria for selection of items as detailed investigations are feasibility and interest rather than suitability to any particular level of teaching. With regard to the suggestions for further work, I have tried to choose topics ranging from those which are easy to undertake to others which are very difficult or have not yet been attempted. Most of them can be completed in a few days or weeks, but a few take up to a year. Other possible lines of investigation are indicated in the main text.

The instructions given with each experiment should be sufficient to allow students to carry out the work from the text, provided that the required materials are made available and the preparatory work properly carried out. The feasibility of each project under local conditions should always be checked before it is attempted. The answers to the questions asked will be found in the main text, to which all the topics are closely tied. One final point—all animals used should be identified to species, otherwise the results in most cases will defy interpretation and have little or no biological meaning.

1 The embryonic development of woodlice

Materials Pregnant female woodlice, preferably *Oniscus asellus* or *Porcellio scaber*, with embryos at various stages of development; CO_2 supply; dissecting microscope; fine mounted needles; a supply of woodlouse Ringer solution in which to culture embryos. This may be made up as follows: to 1 litre of distilled water add 12·0 g NaCl, 1·6 g KCl, 1·6 g $CaCl_2$, 1·6 g $MgCl_2$, and 0·2 g $NaHCO_3$, all anhydrous (or equivalent in hydrated form).

Procedure *a* Anaesthetize with CO_2 a series of ovigerous females. Lift the *oostegites* (flaps) of the brood pouch and remove the embryos using mounted needles. Draw the various stages of development which you can distinguish. .

Do all the members of one brood show the same degree of development? Can you reconstruct the sequence of development?

b Anaesthetize a female with very young embryos. (They look yellow and translucent.) Remove the embryos in Ringer solution and transfer them to clean Ringer. Culture them at room temperature (about 20° C) until they are fully developed. This takes something over 20 days. Every 2 or 3 days note the stage of development and transfer the embryos to fresh Ringer solution in a clean dish.

Does the development of individual embryos confirm the sequence deduced in a?

(Note: Ringer solution is designed to provide the same osmotic concentration, provided by the same principal salts, as the body fluid of the animal or, in this case, as the concentration of the fluid in the brood pouch, which is rather more dilute.)

Further work   Study the effect of temperature on development time. Establish the extent of embryo mortality in the brood pouch by examining brood pouches of full-term young. Check the osmotic concentration of fluid in the brood pouch.

Season   May to October; best in June and July.

Time   Initially 2 to 3 h; half-hour periods at intervals thereafter.

2 Colour change in *Ligia oceanica*

The degree of melanophore expansion determines whether these animals are uniformly light coloured or dark. (The animal shown in plate 2 is intermediate.) Expansion of the melanophores is caused by hormone secretion from the neural tracts between the eye and the brain. Hormone secretion is activated by visual perception of the light level which is, in part, determined by the darkness of the background on which the animal is resting.

Materials   Sixteen *Ligia*, kept for the previous 24 h in tall-sided containers (small plastic buckets, for example) 8 in one with a white background and 8 in one with the floor and sides painted black. Lighting should be diffuse but reasonably strong. The containers should be kept moist and covered by transparent lids. Half the animals in each container should have their eyes thickly covered with a graphite/vaseline mixture to prevent light from entering. Also required: CO_2 supply; hypodermic syringe.

Procedure   *a* Record the number of light and dark individuals in each container.

Do the sighted individuals from dark and light backgrounds differ in colour?

Do the unsighted ones differ? If so, what explanation can you give?

b Use CO_2 to anaesthetize the 4 sighted individuals from the black container, remove the heads, and grind them up in a solid watch glass with 2 cm^3 of sea-water. Allow the solids to settle, then draw off the supernatant liquid in a hypodermic syringe. Making certain that there are no air-bubbles in the syringe, inject a very small quantity of the extract into 2 of the 4 sighted animals from the white container. Inject just to one side of the mid-dorsal line between 2 segments of the thorax. Leave them for about 1 h.

Do you observe any colour change? If so, what has caused it?

Why use extract from animals in the black container rather than the white one?

What do you think would have happened if you had used unsighted animals?

c Boil the remaining extract and inject it into the 2 remaining sighted animals from the white dish.

Is there any effect?

Why boil the extract?

What might be the adaptive value of the response you have been investigating?

Season All year round.

Time 2 to 3 h; this project is best done on a marine course. However, *Ligia* will travel well (except in hot weather) in damp sea-weed, and culture well if moistened regularly with sea-water. Keep them at 10 to 15° C. The experiment will usually show results, even if only 8 animals are available. Check that the act of injection itself does not affect melanophore expansion.

3 The resistance of woodlice to desiccation

Materials *Small* woodlice, about 3 to 4 mm long; the series should consist of *Trichoniscus pusillus*; *Philoscia muscorum* or *O. asellus*; *P. scaber* or another porcellionid; *Armadillidium vulgare* or *A. nasatum*; cobalt thiocyanate $[Co(SCN)_2]$ papers for measuring humidity are also required.

Procedure At room temperature and humidity (about 40%) put the animals into separate dry and empty Petri dishes. Do not allow them to bunch together. Note the time taken for the animals to become moribund—indicated by lack of response to gentle prodding. Check the humidity. Put moribund animals onto a piece of damp filter paper in a closed container, when they will recover. (Note: some woodlice play 'possum', but with practice you should be able to detect this and allow for it.) The time to become moribund is a measure of the resistance to desiccation. Pool the times for at least 5 animals of each species.

What order do the species stand in?

Does this order make sense in view of what you know of the habitat requirements and biology of each species?

Why is it necessary to use small animals, all of the same size?

Further work   Investigate the loss of weight after exposure to air of various humidities.

Season   All year round.

Time   About 3 h except for the most resistant species. To speed up the project, place the Petri dishes in air dried with anhydrous $CaCl_2$.

4 The release of ammonia gas by woodlice

Materials   10 large woodlice; Nessler's reagent for ammonia; $(NH_4)_2SO_4$; 0·1N solution of H_2SO_4; clean test-tubes and pipettes; a balance accurate to 1 mg.

Procedure   The woodlice should be left for 12-h periods (1 overnight, 1 day time) in a 250 cm^3 conical flask stoppered with a cork. A fan of filter paper soaked in 0·1N H_2SO_4 should be pinned to the cork so that it hangs down into the flask. The H_2SO_4 will absorb any ammonia released. The floor of the flask should be covered with damp filter paper so that the woodlice do not dry out and die. Another flask, set up in the same way but without woodlice, should be used as a control. After 12 h remove the filter paper from the cork of the test bottle with clean forceps and soak it for 15 min in 5 cm^3 of distilled water in a sealed test-tube. Shake the tube vigorously, remove the filter paper, and add 0·5 cm^3 of Nessler's reagent and shake the tube again. A brownish-yellow colour indicates that ammonia is present. Treat the filter paper from the control flask and record your observation.

  a Compare the results from animals kept overnight with those of animals kept during the daytime.

Is there any difference? If so, which is darker, and what does this signify?

  b Estimate the actual amounts of ammonia given off by woodlice at night and during the day as follows: take the test solutions just prepared and compare them with a series of standard solutions of known strength. These may be prepared in the following way: weigh out 0·194 g of $(NH_4)_2SO_4$ and dissolve it in 100 cm^3 of 0·1N H_2SO_4. This solution contains 50 mg of NH_3. Take 5 cm^3 of this solution and make it up to 100 cm^3 with distilled water. This new solution contains 25 μg of NH_3 per cm^3. In each of 7 test-tubes put 0·5 cm^3 of Nessler's reagent and add 0·3, 0·6, 0·9, 1·2, 1·5, and 1·8 cm^3 of the 25μg/cm^3 NH_3 solution—the seventh tube receives no NH_3

solution: this is the control. Make each test-tube up to 5.5 cm³ total volume with distilled water. Match the test samples against the known standards and calculate the amounts of ammonia in them. For example, if a test sample matches the standard with 0.6 cm³ of NH_3 solution, it contains 0.6×25 μg of NH_3. Find the total live-weight of animals used and work out the ammonia released in $\mu g/mg$ of animal/h.
How much do the rates vary between night and day?

Note: Nessler's reagent contains mercuric chloride—it is poisonous.

The test is very sensitive and to avoid contamination with stray ammonia, all glassware and other apparatus should be clean and all test-tubes and bottles should be sealed (Parafilm is useful) or stoppered when standing.

Further work Faeces could be examined for ammonia and closer scrutiny of the rhythm of release in relation to activity would be rewarding. Ammonia release in pillbugs needs to be investigated. Since pillbugs lose relatively little water in transpiration perhaps they are unable to lose much ammonia as gas.

Season All year round.

Time 2 to 3 h.

5 The respiratory rate of woodlice

Materials A Dixon-Barcroft respirometer of the type described in detail by Brown, G. D. and Creedy, J. in Experimental Biology Manual (Heinemann, 1970) or that in the Nuffield Advanced Science Handbook "Maintenance of the Organism—a Laboratory Guide", ed. C. F. Stoneman (Penguin, 1970).

Procedure It is best to follow that given in the above books. Compare the rates of oxygen uptake in young juveniles, and in non-pregnant and pregnant females, using 0.5 to 1 g of woodlice on each occasion. (10 large adults of *Oniscus asellus* weigh about 1 g.)

Do these rates differ and, if so, how could you account for the difference?

Study the effect of activity and temperature on respiration rates.

Further work Respirometry is an essential element in the calculation of an energy budget for an individual, population, or trophic level. The calorific equivalent for oxygen uptake can be taken as 21kJ/litre O_2.

Season All year round.

Time 2 to 3 h.

6 The behavioural responses of groups of woodlice

Materials A simple choice chamber as in fig. 36. (The outside edges of the compartments should be sealed to the zinc floor with Plasticine.)

One set should have both compartments transparent and another set should have one side of the apparatus completely darkened. You will also need 10 woodlice (all of the same species) from a moist, *not* wet, culture; $Co(SCN)_2$ papers; anhydrous $CaCl_2$.

Procedure   *a* To study the humidity response use the apparatus with both compartments transparent. Place cotton-wool soaked in water in the container under one compartment, and anhydrous $CaCl_2$ under the other. Check the humidities with $Co(SCN)_2$ papers. Add 5 woodlice to each compartment and plug the holes with Plasticine. After 15 min note the number of woodlice in each compartment. Repeat the procedure with more woodlice.

In which compartment do the animals collect?
What behavioural response is involved?
How might this behaviour aid survival in the wild?

   b Using the apparatus with one side darkened, and with neither water nor $CaCl_2$ below (or if the humidity of the air is very low or the animals very small, with slightly moist cotton-wool below), put 5 animals in each compartment as before, leave them for 15 min, and note their distribution. Repeat the procedure with more woodlice. Answer the same questions as in procedure *a*.

Further work   Test the response of animals kept for the previous 24 h in very dry or very wet conditions. Provide a choice of damp and light against dark and dry to see whether the humidity or light response is stronger. Much will depend upon the environmental conditions experienced by the animals before the test.

Season   All year round.

Time   2 h.

7 The humidity reactions of individual woodlice

Materials   A shallow container with water-tight compartments about 30 cm across: (a plastic *hors d'oeuvres* dish is ideal); a perforated zinc platform; a continuous circle of bunsen tubing; a glass or Perspex lid, the whole assembled as in fig. 36 to form an arena; felt-tip pens; clock with seconds hand; map measurer, a small selection of woodlice; $Co(SCN)_2$ papers, anhydrous $CaCl_2$.

Procedure   Set up the apparatus with anhydrous $CaCl_2$ in one half of the dish and cotton-wool soaked in water in the other, so that half the arena is dry and the other half damp. Leave $Co(SCN)_2$ papers at various points in the arena to check the humidity. Introduce a single woodlouse into the arena through a hole in the centre of the lid and track its progress in the arena by marking its path on the lid with a felt-tip pen. Mark the track at 15-s intervals. Continue for 5 or 10 min;

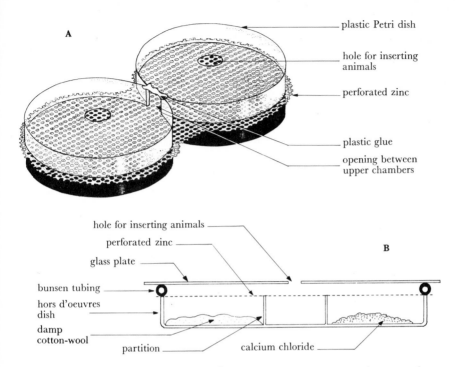

A
- plastic Petri dish
- hole for inserting animals
- perforated zinc
- plastic glue
- opening between upper chambers

hole for inserting animals
perforated zinc
glass plate
B
bunsen tubing
hors d'oeuvres dish
damp cotton-wool
partition
calcium chloride

Fig. 36 Choice chambers for studying locomotory responses to environmental stimuli. **A** a small choice chamber; **B** a large arena (shown in section).

then, with the map measurer, find a, the total distances travelled in the wet and dry halves; b, the total distances in the 2 halves of 'direct' travel—that is, maintaining a straight path or following the curve of the dish. Also calculate for each half the total residence time and the number of sharp turns of more than $90°$.

Do your data suggest that the animal behaves differently in the wet and dry halves?

What behavioural responses are involved?

Can you explain your findings in terms of the humidity requirements of the animal?

Repeat the experiment using different individuals from the same source to discover the extent of individual variation.

Further work The arena provides the opportunity for really detailed studies of individual behavioural responses. Apart from humidity tests, it can be adapted to take hot and cold flowing water and so used as a thermopreferendum apparatus. Phototaxis can be studied, and thigmokinesis. Thigmokinesis can be investigated by glueing Perspex wedges to the lid so that they are positioned close to the wall of the arena, providing areas with a sloping roof and creating artificial 'crevices' with the floor. The response of woodlice to such crevices in various humidities and physiological states can then be investigated. A topic that needs investigation is the comparison of the behaviour of very young and adult animals.

Do newly released woodlice have such well-developed responses to environmental stimuli as their elders?

Season All year round.

Time 2 to 3 h.

8 The humidity reaction of *Porcellio scaber*

Materials A transparent Y-tube of 3-cm bore; two 250-cm^3 flasks with inlet and outlet tubes let into the bungs (inlet tubes should extend to the bottom of the flasks); a supply of compressed air; ten to twenty adult active *P. scaber*; anhydrous $CaCl_2$; $Co(SCN)_2$ papers.

Procedure Without putting anything in the 2 flasks, connect up the inlets to the compressed air supply. Check that the air-flow through the 2 flasks is equal by dipping both outlets into a beaker of water and counting the rate of bubbling. Only a gentle flow is required. Insert a square of $Co(SCN)_2$ paper in each arm of the Y-tube to monitor the humidity and secure them there with a spot of glue in one corner.

Connect up the outlet of one flask to one arm of the Y-tube and the other flask to the other arm of the Y-tube. Make sure that lighting is of equal intensity and direction on both sides of the apparatus and that the apparatus is level. Introduce animals into the stem of the Y one at a time and note the way they turn at the fork. When a clear choice has been made, remove the animal *via* the stem by tilting the apparatus. Change round the flasks and repeat the test. This is the control run to make sure that no unintended stimuli are affecting the behaviour of the animals. Now fill the bottom of one flask with anhydrous $CaCl_2$ and put water in the other. Check the air-flow and connect the 'wet' flask to one arm of the Y-tube and the 'dry' flask to the other arm. Test each animal, then change round the flasks and repeat the test.

Analyse your results statistically using the χ^2 test for 2 variables with expectation (see, for example, Dowdeswell 1959 p. 118). The null hypothesis is that an animal is as likely to choose one arm of the Y as the other. First, analyse the results for the control run to ensure that there is no undetected bias towards one arm or the other. Then, if the control results conform to the null hypothesis (as they should if the sample is big enough) calculate χ^2 for the test results. If several sets of apparatus are being used the results can be pooled, providing the control run in each case is satisfactory.

Does your analysis show that the test animals have a statistically significant preference for wet air or dry air?

How do you interpret the behaviour you have observed?

Further work Humidity responses can be examined in detail by filling the flasks with saturated solutions of salts which impart known relative humidities to air passed through them. (An aquarium aerator stone should be used to provide a slow stream of very small bubbles through the solution.) Solutions should be made up as follows: barely saturate boiling water with the salt and allow the solution to cool partially. Add a little more of the salt. After complete cooling add considerably more of the salt and let the solution stand for several days before use.

Useful saturated solutions (with the % relative humidity at 20° C) are as follows: LiCl (12·5); $MgCl_2 . 6H_2O$ (33·0); KNO_2 (48·5); NaCl (76·0); KCl (85·0); KNO_3 (93·5). Many others will be found in a paper by Winston, P. W. and Bates, D. H. (1960) in *Ecology* **41**, 232–237.

Olfactory responses of *P. scaber* to woodlice of the same or other species (contained in one of the flasks) can be tested also. Likewise, you can investigate reactions to NH_3, CO_2, and other gases at various concentrations. Response to uropod secretions could be looked at in detail.

Season All year round.

Time 3 h minimum.

9 The humidity requirements of woodlice in the field

Materials A study area about the size of a tennis court or larger in grassland or on waste ground where woodlice are known to be plentiful; a series of about 20 shelter sites, which may be old tiles, pieces of paving, bricks, cardboard, loose turves, or wood; unseasoned, treated, or painted timber is unsuitable.

Procedure The shelter sites should be established some weeks before the project is begun. They should be shielded from direct sun (with straw, for example) but so placed that they will dry out when required to do so if protected from rain. In a drought, artificial watering may be necessary and checks on humidity can be made with $Co(SCN)_2$ papers mounted in a probe to prevent contact with wet surfaces.

Count the numbers of woodlice (identified to species) under each site at daily intervals for a week, maintaining moisture conditions at a fairly constant level. Establish a mean figure for each shelter. Now allow half the shelters to dry out by protecting them from rain or, in dry weather, by ceasing to water them. Maintain the other shelters at the same humidity as before. After a week, count the animals under each shelter daily for a week and establish means for wet and dry shelters.

What has been the effect of drying out the shelters on woodlouse numbers? Has each species been affected to the same degree?

Are there statistically significant differences in numbers of each species under the wet and the dry shelters?

Can you explain any difference between species which you have found?

Season May to August.

Time Roughly 1 h daily for 2 separate weeks.

10 Does a woodlouse have an habitual shelter site?

Materials *a* Vivarium 30 × 60 cm with a moist floor and 4 identical shelters; marking paint; 4 specimens of *P. scaber*; *b* an area of woodland, grassland, or waste ground several square metres in area. The shelter sites should be established some weeks before the project is begun.

Procedure *a* Mark each specimen with a different colour. Marking techniques are given in Chapter 9. Set the woodlice free in the vivarium. Inspect them each day until the marks are lost.

Does each animal spend all its time under one shelter? If not, what is the residence pattern?

b Mark the animals under each shelter with a separate colour. Try to mark about 100 animals altogether. If the loss-rate of marked animals is so high that all are gone in a few days, try maintaining more shelters. Allow the paint to dry, then put each animal back under the shelter from which it was collected. This is its 'home site'. Each home site should be marked with the colour of the animals put under it. Re-examine the shelters daily, recording *a* the numbers of marked specimens still in their home sites; *b* the number of marked specimens which have changed shelters. In both cases add up totals for all the shelter sites and then, for each date, calculate the total number of animals still in their home sites as a percentage of the total marked animals under all the shelters (the home site percentage). Plot this percentage against time.

Do your data suggest that each woodlouse is strongly attached to a particular shelter site?

If woodlice habitually return to the same shelter after nocturnal activity, the home site percentage will initially be very high, declining only slowly. With complete fidelity the percentage will be 100. A low level of fidelity will give a rapid initial drop followed by a levelling to equilibrium when animals start to return to the original site after sheltering elsewhere. Where the area is restricted and the choice of shelter sites is small (as in vivaria) this equilibrium may be reached quite quickly. A lack of equilibrium indicates that a very wide choice of sites is available or that animals are avoiding their original home sites.

Further work By varying percentage rates of loss and return from and to the home site, construct a numerical model which will match the changes in home site percentage which you have observed.

Season April to October in mild weather.

Time Daily or every other day for up to 2 weeks.

1 Estimating population size

Material Well established shelter sites, as described previously, distributed over the study area; marking paint or dye; woodlice (*P. scaber* is most suitable because its rough cuticle holds paint well.)

Procedure Capture as many woodlice as possible, and mark them. Allow the paint to dry, remove any injured specimens, and release the rest by scattering them about the site (preferably at dusk). After 4 days of unexceptional weather count the numbers of marked and unmarked animals under each shelter. Assuming that there has been random mixing of marked and unmarked individuals, the total population size is given by:

$$\frac{\text{no. collected in 1st sample} \times \text{no. collected in 2nd sample}}{\text{no. of marked individuals in 2nd sample}}$$

What is the estimated size of this population?

Why recapture after four days rather than one?

Some of the paint marks will flake off and marked individuals be mistaken for unmarked ones. What effect will this error have on your population estimate?

What other sources of error are involved?

Season April to October in mild weather.

Time 2 to 3 h for 2 days.

2 The rate of movement from a central release point

Materials A large and well established shelter site in the centre of a suitable field study area; satellite shelters distributed at set distances along 4 or more radii from the central release point; marking paint.

Procedure Mark the woodlice under the central shelter for several days running, using a different colour for each day. Count the number of marked individuals under the satellite shelters at intervals (1, 2, or

3 days) thereafter for 10 to 15 days. Calculate the maximum and mean distances moved by the marked individuals at each time interval after release. (The minimum distance moved will probably remain at 0.)

How rapidly do they disperse? Is there a great difference between maximum and mean dispersal rates?

Divide the area into quadrants on a map. Add up the total numbers of individuals recaptured in each quadrant.

Is there any directional bias? If so, what could account for it?

Further work Collect up a large number of woodlice and stage a mass release at the central point. Compare the results with those you have just obtained. Does the enforced aggregation lead to an artificially high rate of dispersal? Test the results of this and the directional bias statistically.

Season April to October in mild weather.

Time 1 h for 10 to 15 days.

13 The degree of aggregation of woodlice

Materials Pooters; widgers; string; skewers; flower-pot labels; turf cutter; bread knives; a uniform piece of grassland in which the larger woodlice are known to be common.

Procedure Mark out an undisturbed area of 1 m². Keeping off the area, map the vegetation within it. Mark the edges every 20 cm and with skewers and string divide it up into 25 small squares. Working towards the centre, and avoiding all trampling of the material in the square, cut out and remove the marked out small squares using the turf cutter. Cut the turves at a depth of 5 cm of soil. As it is removed, give each square a label to show its position in the large square. Shave off the growing grass and shake each turf upside down over a white sheet or tray. Poot up animals as they appear. Search carefully among the bases of the grass stems. Collect up the woodlice from each small square in separate tubes and take them back to the laboratory for counting and identification. Before leaving the site, replace all the turves.

Treating each species separately plot out the number of individuals in each small square on the map showing the main vegetational features of the area sampled. Select limits for high, medium, and low densities. Assign each density a colour, and colour in the small squares on the map accordingly.

Does any pattern emerge? If so, how big are the aggregations and do they seem to be associated with vegetational topography such as tussocks and so on?

Calculate the mean and variance for each species.

Does the variance divided by the mean come to >1, 1, or <1 (indicating an aggregated, random, or regular distribution respectively)?
What might be the biological significance of the distribution you observe?
Do all species show the same pattern?
Further work   Check the efficiency of hand-sorting. Test data for goodness of fit with Poisson and normal distributions.
Season   April to October.
Time   2 to 3 h for 2 to 3 days, depending on man-power.

4 The habitat ranges of woodlice

Materials   5 Tullgren funnels (fig. 35); bread knives (and first aid kit); large polythene bags.
Procedure   Take sets of 5 cores between 16 and 30 cm diameter of soil and litter to a depth of 5 cm of soil from as wide a variety of sites as possible. These should be selected from the following: ungrazed grassland, heavily grazed pasture, arable land, coniferous woodland, and deciduous woodland. Sites on both acid and basic soils should be looked at in the case of grassland and forest. Other habitats which could be added are marshland, acid heath, and maritime grassland. The cores should be taken within a few metres of each other in a uniform area, the details of this site being carefully recorded on an Isopod Survey Scheme card (pp. 132–133). Extract all the woodlice using the Tullgren funnels and identify them to species. Estimate the pH of the extracted cores. Pool numbers of individuals of each species and the pH of the soil.
Which habitat supports a, the most species? b, the highest total of individuals?
Is there any correlation between pH of the soil at each site and a, number of species b, total individuals? (Investigate this by plotting a graph or calculating the correlation coefficient.)
What are the effects of cultivation and grazing on a, species diversity b, the numbers of each species? Can you suggest any explanation for these effects?
Further work   Analyse the relation between Ca content and numbers of each species on each site. Determine the Ca content of the exoskeleton for each species, calculating correlation coefficients between this and Ca content of sites. Convert density figures to biomass and analyse them on this basis.

Calcium content and pH are, of course, only 2 of a larger number of variables to be considered. Others include water content, total soil organic matter, $C:N$ ratio (see Dowdeswell, 1959, for details of treatment).

Season All year round, but beware of high density levels in any species because of sampling shortly after release of young.

Time 3-h periods once a week for some weeks.

The project can also be carried out using the larger woodlice by recording the number of woodlice collected by hand-sorting on the various sites over standard time periods. However, results could be badly affected because woodlice are easier to see in some habitats than they are in others.

15 Local names, folklore, and woodlice

Being common everyday creatures, woodlice have been given local names in many parts of the country. Sometimes a single name is used for them all but, very often, different types have different names—slater and pillbug, for example.

Procedure *a* Make a collection of a series of different types of woodlice in tubes of alcohol and ask people, particularly those brought up in country districts, what names they give to them. Make a note of where each person lived as a child and then plot on a map the distribution of the names they have given.

Can you find regional differences in names? If so, do the names change abruptly from one area to another, or are there related terms in adjacent areas?

Can you locate the geographical origin of the following terms: sowbug (now the usual term for a woodlouse in the United States); parson's pigs; tiggy hogs; grammar sows; granfers?

There are many other names (see Webb and Sillem 1906, or papers by W. A. Collinge in the North-Western Naturalist up to 1947).

Can you find (or think of) any explanation for the predominance of names suggesting a likeness to pigs, or association with the Church?

b Woodlice were sometimes used to cure various ailments. In Herefordshire, tying a black silk bag containing 7 to 9 live woodlice around a baby's neck was supposed to make teething less painful. Can you unearth any similar cures or remedies, either from talking to old people or by searching the older literature? A good source of references is Dr W. T. Fernie's 'Animal Simples' published by J. Wright and Company in 1899.

Further lines of investigation

Morphology

1 How do woodlice walk? Compare the leg action of *O. asellus* with that of the marine isopods *Idotea*.

2 Compare the leg action of *O. asellus* with that of the centipede *Lithobius*, an iulid millipede, and a cockroach.

3 Investigate the occurrence of ocelli and true compound eyes in the various families of woodlice. What significance might your results have on the study of evolutionary relationships of woodlice?

Physiology

1 Study the uptake of water at high humidities ($>95\%$ R.H.) and the effect on transpiration of waterproofing the cuticle and/or pseudo-tracheae.

2 How does the water content of the faeces vary in relation to the previous history of the animal?

3 Study the survival of woodlice in fresh-water and sea-water at various levels of temperature and oxygenation.

4 Investigate pleopod beat in *L. oceanica* and other woodlice when put into water. What bearing might pleopod function have on theories of the ancestry of woodlice?

5 Analyse the secretions of the uropod and lateral plate glands.

6 Why do woodlice eat filter paper?

7 What enzymes are produced by the hepatopancreas?

8 Do woodlice have an extensive gut flora? Compare it with the oniscomorph millipede *Glomeris* by examining gut structure.

9 Attempt to transplant androgenic glands from mature males into mature females and study the effects on the sex organs.

10 What is the nature of the cuticular pigments?

Behaviour

1 What is the response of a woodlouse when flooded with water from a pipette?

2 Study the water distribution system on the underside of woodlice using dyed water.

3 Observe and describe the mating behaviour of woodlice.

4 Study the defence reactions of woodlice to predators.

5 What is the response of fertilized female fly parasites to uropod gland secretions?

Genetics

1 Examine the incidence of colour forms in various species, habitats, and regions at different times of year.

2 Study the incidence of single-sex broods.

3 Investigate sex-ratios in diploid and triploid populations of *T. pusillus*.

Food, predators, and parasites

1 Compare faeces production of animals fed on litter of various ages and types (deciduous, coniferous, grass, and so on).

2 How extensive is coprophagy (eating their own faeces) among woodlice?

3 Compare the effects of woodlice on the breakdown of litter.

4 Look for predation on woodlice by observation after dark.

Population ecology

1 Calculate the regression of embryo-number on female head-width and establish the relationship between live-weight and head-width for a variety of species.

2 Follow changes in the size structure and numbers of a population over a year by regular sampling. Use the data to calculate annual recruitment, mortality, and change in biomass.

3 Investigate the effects of fly parasites on *P. scaber* populations.

4 Study the microclimate of the shelters in which woodlice live.

5 Investigate the effect of wind on woodlouse activity.

Distribution

1 Study the abundance of *A. vulgare* in relation to Ca content and pH of the soil, working inland from one of the coastal sites where it occurs on soils derived from non-basic rocks, as those along the North Cornish coast.

Appendix

The study of isopod distribution in Britain

Although, for a few woodlice species, we have a fairly good idea of geographical distribution and the range of habitat occupied, for the great majority of British species we know exceedingly little. It is true that at the turn of the century there was a great deal of interest in woodlice and much active collecting, but unfortunately many of the early records are unreliable, mainly because the aids to identification were inadequate. Furthermore, these early records give little indication of habitat, and in any case refer to a landscape and pattern of land-use now somewhat changed. What is needed is accurate knowledge of present-day distribution and habitat range for each British species.

Such knowledge can be gained only by systematic recording which should have the following aims:

1 to clarify the status of the British species as native, naturalized, or alien;
2 to show which habitats are occupied by which species, as a step towards finding out the factors limiting the range of habitats occupied by each species;
3 to provide sufficient knowledge of distribution so that future changes can be detected;
4 to indicate which species are genuinely rare and in need of conservation.

With regard to this last point we have, at the moment, a large number of seemingly rare or very rare species, but so little recording has been done in recent years that it is difficult to counter the argument that this rarity is more apparent than real and thus to make any effective case for conservation. One of the 6 known sites for *Haplophthalmus mengei* has recently been lost because of this lack of information.

With these problems in mind a Survey Scheme has been organized to collect information about intertidal, freshwater, and terrestrial isopods in the British Isles. The basis of the Survey is a card on which locality and habitat details can be recorded wherever isopods are

131

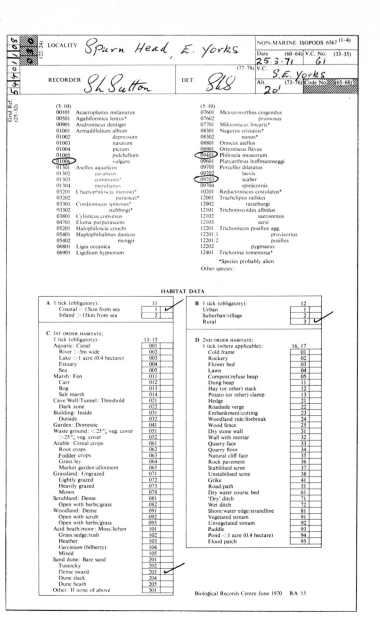

(22–24)	**LOCALITY** *Spurn Head, E. Yorks*	**NON-MARINE ISOPODS 6567** (1–4)	
		Date (60–64)	V.C. No. (33–35)
		25.3.71	*61*
		(77–79) V.C. *S.E. Yorks*	
Grid Ref. (25–32) *5 4 0 1 1 0 8*	**RECORDER** *S.L. Sutton* DET. *SLS*	Alt. (73–76) Code No. (65–68) *20'*	

(5–10)		(5–10)	
00101	Acaeroplastes melanurus	07601	Metoponorthus cingendus
00501	Agabiformius lentus*	07602	pruinosus
00901	Androniscus dentiger	07701	Miktoniscus linearis*
01001	Armadillidium album	08301	Nagurus cristatus*
01002	depressum	08302	nanus*
01003	nasatum	08801	Oniscus asellus
01004	pictum	08901	Oritoniscus flavus
01005	pulchellum	09401	Philoscia muscorum
01006	vulgare	09601	Platyarthrus hoffmannseggi
01301	Asellus aquaticus	09701	Porcellio dilatatus
01302	cavaticus	09702	laevis
01303	communis*	09703	scaber
01304	meridianus	09704	spinicornis
03201	Chaetophiloscia mecusei*	10201	Reductoniscus costulatus*
03202	patiencei*	12001	Trachelipus rathkei
03301	Cordioniscus spinosus*	12002	ratzeburgi
03302	stebbingi*	12101	Trichoniscoides albidus
03801	Cylisticus convexus	12102	saeroeensis
04701	Eluma purpurascens	12103	sarsi
05201	Halophiloscia couchi	12201	Trichoniscus pusillus agg.
05401	Haplophthalmus danicus	12201.1	provisorius
05402	mengei	12201.2	pusillus
06801	Ligia oceanica	12202	pygmaeus
06901	Ligidium hypnorum	12401	Trichorina tomentosa*

*Species probably alien

Other species:

HABITAT DATA

A 1 tick (obligatory):	11	
Coastal <15km from sea	1	✔
Inland >15km from sea	2	

B 1 tick (obligatory):	12	
Urban	1	
Suburban/village	2	
Rural	3	✔

C 1ST ORDER HABITATS;		
1 tick (obligatory):	13–15	
Aquatic: Canal	001	
River >5m wide	002	
Lake >1 acre (0.4 hectare)	003	
Estuary	004	
Sea	005	
Marsh: Fen	011	
Carr	012	
Bog	013	
Salt marsh	014	
Cave/Well/Tunnel: Threshold	021	
Dark zone	022	
Building: Inside	031	
Outside	032	
Garden: Domestic	041	
Waste ground: <25% veg. cover	051	
>25% veg. cover	052	
Arable: Cereal crops	061	
Root crops	062	
Fodder crops	063	
Grass ley	064	
Market garden/allotment	065	
Grassland: Ungrazed	071	
Lightly grazed	072	
Heavily grazed	073	
Mown	074	
Scrubland: Dense	081	
Open with herbs/grass	082	
Woodland: Dense	091	
Open with scrub	092	
Open with herbs/grass	093	
Acid heath/moor: Moss/lichen	101	
Grass/sedge/rush	102	
Heather	103	
Vaccinium (bilberry)	104	
Mixed	105	
Sand dune: Bare sand	201	
Tussocky	202	
Dense sward	203	✔
Dune slack	204	
Dune heath	205	
Other: If none of above	301	

D 2ND ORDER HABITATS;		
1 tick (where applicable):	16, 17	
Cold frame	01	
Rockery	02	
Flower bed	03	
Lawn	04	
Compost/refuse heap	05	
Dung heap	11	
Hay (or other) stack	12	
Potato (or other) clamp	13	
Hedge	21	
Roadside verge	22	
Embankment/cutting	23	
Woodland ride/firebreak	24	
Wood fence	25	
Dry stone wall	31	
Wall with mortar	32	
Quarry face	33	
Quarry floor	34	
Natural cliff face	35	
Rock pavement	36	
Stabilised scree	37	
Unstabilised scree	38	
Grike	41	
Road/path	51	
Dry water course bed	61	
'Dry' ditch	71	
Wet ditch	72	
Shore/water edge/strandline	81	
Vegetated stream	91	
Unvegetated stream	92	
Puddle	93	
Pond <1 acre (0.4 hectare)	94	
Flood patch	95	

Biological Records Centre June 1970   RA 15

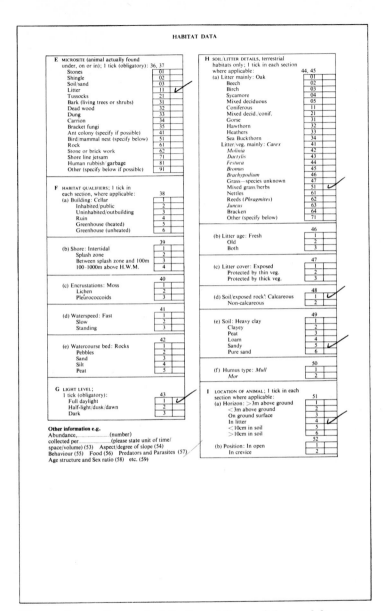

HABITAT DATA

E MICROSITE (animal actually found under, on or in); 1 tick (obligatory): 36, 37

Stones	01
Shingle	02
Soil/sand	03
Litter	11 ✓
Tussocks	21
Bark (living trees or shrubs)	31
Dead wood	32
Dung	33
Carrion	34
Bracket fungi	35
Ant colony (specify if possible)	41
Bird/mammal nest (specify below)	51
Rock	61
Stone or brick work	62
Shore line jetsam	71
Human rubbish/garbage	81
Other (specify below if possible)	91

F HABITAT QUALIFIERS; 1 tick in each section, where applicable: 38

(a) Building: Cellar	1
Inhabited/public	2
Uninhabited/outbuilding	3
Ruin	4
Greenhouse (heated)	5
Greenhouse (unheated)	6

	39
(b) Shore: Intertidal	1
Splash zone	2
Between splash zone and 100m	3
100-1000m above H.W.M.	4

	40
(c) Encrustations: Moss	1
Lichen	2
Pleurococcoids	3

	41
(d) Waterspeed: Fast	1
Slow	2
Standing	3

	42
(e) Watercourse bed: Rocks	1
Pebbles	2
Sand	3
Silt	4
Peat	5

G LIGHT LEVEL; 1 tick (obligatory): 43

Full daylight	1 ✓
Half-light/dusk/dawn	2
Dark	3

Other information e.g.
Abundance,_____(number)
collected per_____(please state unit of time/
space/volume) (53)   Aspect/degree of slope (54)
Behaviour (55)   Food (56)   Predators and Parasites (57)
Age structure and Sex ratio (58)   etc. (59)

H SOIL/LITTER DETAILS, terrestrial habitats only; 1 tick in each section where applicable: 44, 45

(a) Litter mainly: Oak	01
Beech	02
Birch	03
Sycamore	04
Mixed deciduous	05
Coniferous	11
Mixed decid./conif.	21
Gorse	31
Hawthorn	32
Heathers	33
Sea Buckthorn	34
Litter/veg. mainly: *Carex*	41
Molinia	42
Dactylis	43
Festuca	44
Bromus	45
Brachypodium	46
Grass—species unknown	47
Mixed grass/herbs	51 ✓
Nettles	61
Reeds (*Phragmites*)	62
Juncus	63
Bracken	64
Other (specify below)	71

(b) Litter age: Fresh	46
	1
Old	2
Both	3

(c) Litter cover: Exposed	47
	1
Protected by thin veg.	2
Protected by thick veg.	3

(d) Soil/exposed rock: Calcareous	48
	1 ✓
Non-calcareous	2

(e) Soil: Heavy clay	49
	1
Clayey	2
Peat	3
Loam	4
Sandy	5 ✓
Pure sand	6

(f) Humus type: *Mull*	50
	1
Mor	2

I LOCATION OF ANIMAL; 1 tick in each section where applicable: 51

(a) Horizon: >3m above ground	1
<3m above ground	2
On ground surface	3
In litter	4 ✓
<10cm in soil	5
>10cm in soil	6

(b) Position: In open	52
	1
In crevice	2

Fig. 37 A recording card of the Isopod Survey Scheme; *left*, page 1; *above*, the reverse side. The card is shown about ¾ actual size.

collected, one card serving the marine families and another the fresh-water and terrestrial species. Participants in the Scheme fill in a card recording all the species found together in one 'microsite'. This might be a piece of driftwood, the walls of a cellar, or the undersides of bricks on a piece of waste ground (see fig. 37 for card details). A different card is used for each collection so that each card refers to the species found together at a particular microsite, in a particular locality, at a particular time. (Detailed guidance on how to fill out cards is given to all participants.) The card is so arranged that the information given can be coded and punched on to another card, which can then be used to feed data to a computer for analysis. In this way, data from many thousands of cards can be handled. Analysis of locality data includes the printing of distribution maps, while analysis of the habitat data should clarify such questions as: How extensive is the overlap in habitat range of *O. asellus*, *P. scaber*, and *P. muscorum*? Is *Metoponorthus pruinosus* primarily a farmyard species? Which habitats support the richest variety of woodlice? Is *Platyarthrus hoffmannseggi* restricted to calcareous soils as well as to ants' nests? With this kind of background information, detailed studies can be carried out to identify the factors determining habitat range.

The scheme was started in 1968 and is run by a small group of people with the assistance of the Biological Records Centre of the Nature Conservancy, which also produces maps of distribution. Initially, the aim of the Survey is to sample all the major habitats in each of the $(50 km)^2$ areas of the National Grid into which the British Isles is divided. Small expeditions to the more remote parts of Scotland, Wales, and Ireland are sent out to ensure adequate coverage of these areas so as to prevent the maps reflecting the distribution of collectors rather than of woodlice. When this first phase is over, the distribution of the more restricted species will be looked at in greater detail.

To be successful, a scheme of this kind must maintain a high level of accuracy when identifying species. To ensure this, each participant in the Scheme initially sends all his specimens to a panel of experienced collectors for checking, and these are returned to him so that he can build up his own voucher collection.

Inevitably, the Scheme involves the collecting of rare species, but only in small numbers. Further details can be obtained from the Biological Records Centre of the Nature Conservancy, Monks Woods Experimental Station, Abbots Ripton, Huntingdonshire.

Similar schemes can be run privately, of course, in order to develop skill in analysing habitats and identifying the woodlice found, using the key in Chapter 8.

BIBLIOGRAPHY

Adamkewicz, S. Laura   1969   Colour polymorphism in the land isopod *Armadillidium nasatum. Heredity, Lond.* **24,** 249–264.

Bailey, N. T. J.   1959   Statistical methods in biology. English Universities Press, London.

Bedding, R. A.   1965   Parasitism of British terrestrial Isopoda by Diptera. Unpublished Ph.D thesis, Imperial College of Science and Technology, London.

Beirne, B. P.   1952   The origin and history of the British fauna. Methuen, London.

Carthy, J. D.   1958   An introduction to the behaviour of invertebrates. Allen & Unwin, London.

Charniaux-cotton, H.   1960   Sex determination; *in* Physiology in Crustacea Vol. 1. ed. T. H. Waterman. Academic Press, New York.

Clark, W. T.   1970   Predation by the toad *Bufo valliceps* and regulation of density of prey populations of *A. vulgare.* Unpublished Ph.D. thesis, University of Texas at Austin, U.S.A.

Cloudsley-Thompson, J. L.   1952   Diurnal rhythms in woodlice. *J. exp. Biol.* **29,** 295–303.

Cloudsley-Thompson, J. L.   1958   The effect of wind upon the nocturnal emergence of woodlice and other terrestrial arthropods II. *Entomologist's mon. Mag.* **94,** 184–5.

Cloudsley-Thompson, J. L.   1971   The water and temperature relations of woodlice (Isopoda: Oniscoidea). Merrow Publishing Co., Watford, Herts.

Den Boer, P. J.   1961   The ecological significance of activity patterns in the woodlouse *Porcellio scaber* Latr. (Isopoda). *Archs. néerl. Zool.* **14,** 283–409.

| Dowdeswell, W. H. | 1959 | Practical Animal Ecology. Methuen, London. |

Edney, E. B. 1953 The woodlice of Great Britain and Ireland —a concise systematic monograph. *Proc. Linn. Soc. Lond.* **164**, 49–98; (reprinted in shortened form in 1954 as Linnean Society Synopsis of the British Fauna No. 9, British Woodlice).

Edney, E. B. 1957 The water relations of terrestrial arthropods. Cambridge Monographs in Experimental Biology No. 5. Cambridge University Press.

Edney, E. B. 1968 Transition from water to land in Isopod Crustaceans. *Am. Zool.* **8**, 309–326.

Foster, N. H. 1911 On two exotic species of woodlice found in Ireland. *Irish Naturalist* **20**, 154–156.

Friedlander, C. P. 1963 Thigmokinesis in woodlice. *Anim. Behav.* **12**, 164–174.

Goodrich, A. C. 1939 The origin and fate of the endoderm elements in the embryogeny of *Porcellio laevis* Latr. and *Armadillidium vulgare* Latr. (Isopoda). *J. Morph.* **64**, 401–430.

Gorvett, H. 1956 Tegumental glands and terrestrial life in woodlice. *Proc. zool. Soc. Lond.* **126**, 291–314.

Gorvett, H. & Taylor, J. C. 1960 A further note on tegumental glands in woodlice. *Proc. zool. Soc. Lond.* **133**, 653–655.

Gruner, H. E. 1965 1966 Krebstiere oder Crustacea V: Isopoda. VEB Gustav Fischer Verlag, Jena, 2 vols.

Harding, P. T. 1968 Notes on the biology and distribution of *Armadillidium album* Dollfus (Crustacea: Isopoda, Oniscoidea) in the British Isles. *Entomologist's mon. Mag.* **104**, 269–272.

Hartenstein, R. 1968 Nitrogen metabolism in the terrestrial isopod *Oniscus asellus. Am. Zool.* **8**, 507–519.

Heeley, W. 1941 Observations on the life histories of some terrestrial isopods. *Proc. zool. Soc. Lond.* B **3**, 79–149.

Holdich, D. & Ratcliffe, N. A. 1970 A light and electron microscope study of the hindgut of the herbivorous isopod

Dynamene bidentata (Crustacea: Peracarida). *Z. zellforsch. mikrosk. Anat.* **111**, 209–227.

Holthuis, L. B. 1947 On a small collection of Isopod Crustacea from the greenhouses of the Royal Botanic Gardens, Kew. *Ann. Mag. nat. Hist. (Ser. 11)* **13**, 122–137.

Howard, H. W. 1962 The Genetics of *Armadillidium vulgare* Latr. V; factors for body colour. *J. Genet.* **58**, 29–38.

Hubbell, S. P., Sikora, A., & Paris, O. H. 1965 Radiotracer, gravimetric and calorimetric studies of ingestion and assimilation rates of an isopod. *Hlth Phys.* **11**, 1485–1501.

Hussey, N. W., Read, W. H., & Hesling, J. J. 1969 The Pests of Protected Cultivation. Edward Arnold, London.

Ing, B. 1967 Myxomycetes as food for other organisms. *Proc. S. Lond. ent. nat. Hist. Soc.* 1967, 18–23.

Jans, D. & Ross, K. F. A. 1963 A histological study of the peripheral receptors in the thorax of land isopods, with special reference to the location of possible hygroreceptors. *Q. Jl. microsc. Sci.* **104**, 337–350.

Kuenen, D. J. 1959 Excretion and water balance in some land-isopods. *Entomologia exp. appl.* **2**, 287–294.

Kuenen, D. J. & Nooteboom, H. P. 1963 Olfactory orientation in some land isopods (Oniscoidea, Crustacea). *Entomologia exp. appl.* **6**, 133–142.

Lattin, G. de 1954 Zur populationsgenetik geschlechtsbeeinflussender Farbfaktoren bei Porcellioniden (Crust. Isop.) *Caryologia* **6**, (Suppl.), 883–888.

Legrand, J. J., Juchault, P., Mocquard, J. P., & Noulin, G. 1968 Contribution a l'étude du contrôle neuro-humoral de la physiologie sexuelle mâle chez les Crustaces isopodes terrestres. *Ann. Embr. Morph.* **1**, 87–105.

Lindquist, O. V. 1968 Water regulation in terrestrial isopods, with comments on their behaviour in a

		stimulus gradient. *Ann. Zool. Fenn.* **5,** 279–311.
Lindroth, C. H.	1957	The faunal connections between Europe and North America. Wiley, New York.
McWhinnie, M. A. & Sweeney, H. M.	1955	The demonstration of two chromatophoro-tropically active substances in the land isopod *Trachelipus rathkei. Biol. Bull. mar. biol. lab. Woods Hole, Mass.* **108,** 160–174.
Palmén, E.	1958	The role of European species in the New-foundland fauna of Chilopods, Diplopods and terrestrial Isopods. *Proceedings of the Tenth International Congress of Entomology* **1,** 899–902.
Paris, O. H.	1963	The Ecology of *Armadillidium vulgare* (Isopoda: Oniscoidea) in California grass-land: food, enemies and weather. *Ecol. Monogr.* **33,** 1–22.
Paris, O. H. & Sikora, Anne	1965	Radiotracer demonstration of isopod herbi-vory. *Ecology* **46,** 729–734.
Paris, O. H. & Sikora, Anne	1967	Radiotracer analysis of the trophic dyna-mics of natural isopod populations; (*in* Secondary Productivity of Terrestrial Eco-systems, ed. K. Petrusewicz). Warsaw Pan-stowe Wydawnicto Naukoe.
Petrusewicz, K. & Macfayden, A.	1970	IBP Handbook No. 13; Productivity of Terrestrial Animals, Principles and Meth-ods. Blackwell, Oxford.
Phillipson, J. & Watson, J.	1965	Respiratory metabolism of the terrestrial isopod *Oniscus asellus* L. *Oikos* **16,** 78–87.
Rudge, M. R.	1968	The food of the common shrew *Sorex araneus* L. (Insectivora-Soricidae) in Brit-ain. *J. anim. Ecol.* **37,** 565–581.
Russell-Hunter, W. D.	1969	A biology of higher invertebrates. Mac-millan Co., London.
Saito, S.	1965	Structure and energetics of the population of *Ligidium japonicum* (Isopoda) in a warm temperate forest ecosystem. *Jap. J. Ecol.* **15,** 47–55.

Schmitz, E. H.,   1969   Digestive anatomy of terrestrial Isopoda:
&
Schultz, T. W.

Armadillidium vulgare and *Armadillidium nasatum. Am. Midl. Nat.* **82**, 163–181.

Schmöltzer, K.   1965   Ordnung Isopoda (Landasseln) Akademie-Verlag, Berlin.

Sheppard, E. M.   1968   *Trichoniscoides saeroeensis* Lohmander, an isopod new to the British fauna. *Trans. Cave Research Group of G.B.* **10**, 135–138.

Smith, W. J.,   1969   Structural adaptations for ion and water
Witkus, E. R., &
Grillo, R. S.

transport in the hind gut of the woodlouse *Oniscus asellus. J. Cell Biol.* **43**, 135–136a.

Snedecor, G. W.   1967   Statistical Methods. Iowa State University
& Cochran, W. G.

Press (6th edn.).

Southwood,   1966   Ecological Methods. Methuen, London.
T. R. E.

Spencer, J. O. &   1954   The absorption of water by woodlice.
Edney, E. B.

J. exp. Biol. **31**, 491–496.

Stachurski, A.   1968   Emigration and mortality rates and the food-shelter conditions of *Ligidium hypnorum* L. (Isopoda). *Ekol. pol. Seria A* **16**, 445–459.

Standen, V.   1970   The life history of *Trichoniscus pusillus pusillus* (Crustacea: Isopoda) *J. Zool., Lond.* **161**, 461–470.

Stevenson,   1961   Polyphenol oxidase in the tegumental glands
J. Ross

in relation to the molting cycle of the isopod Crustacean *Armadillidium vulgare. Biol. Bull. mar. biol. Lab., Woods Hole, Mass.* **121**, 554–560.

Stevenson,   1967   Mucopolysaccharide glands in the isopod
J. Ross &
Murphy, J. C.

crustacean *Armadillidium vulgare. Trans. Am. microsc. Soc.* **86**, 50–57.

Sutton, S. L.   1966   The Ecology of Isopod Populations in Grassland. Unpublished D.Phil. thesis, Oxford University.

Sutton, S. L.   1968   The population dynamics of *Trichoniscus pusillus* and *Philoscia muscorum* (Crustacea,

		Oniscoidea) in limestone grassland. *J. anim. Ecol.* **37**, 425–444.
Sutton, S. L.	1969	The study of woodlice. *Proc. Brit. ent. Soc.* **1**, 71–75.
Sutton, S. L.	1970	Predation on woodlice—an investigation using the precipitin test. *Entomologia exp. appl.* **13**, 279–285.
Sutton, S. L.	1970	Growth patterns in *Trichoniscus pusillus* and *Philoscia muscorum* (Crustacea: Oniscoidea). *Pedobiologia* **10**, 434–441.
Thompson, W. R.	1934	The tachinid parasites of woodlice. *Parasitology* **26**, 378–448.
Wallwork, J. A.	1970	Ecology of Soil Animals. McGraw-Hill, London.
Warburg, M. R.	1965	Water relations and internal body temperature of isopods from mesic and xeric habitats. *Physiol. Zoöl.* **38**, 99–109.
Webb, W. M. & Sillem, C.	1906	The British Woodlice. Duckworth, London.
Wieser, W.	1965	Electrophoretic studies on blood proteins in an ecological series of isopod and amphipod species. *J. mar. biol. Ass. U.K.* **45**, 507–523.
Wieser, W.	1966	Copper and the role of isopods in the degradation of organic matter. *Science, N.Y.* **153**, 67–69.
Wieser, W., Schweizer, G., & Hartenstein, R.	1969	Patterns of release of gaseous ammonia by terrestrial arthropods. *Oecologia (Berl.)* **3**, 390–400.
Vandel, A.	1940	La parthénogenèse geographique IV. Polyploidie et distribution geographique. *Bull. biol. Fr. Belg.* **74**, 94–100.
Vandel, A.	1960 1962	Isopodes Terrestres. *Faune de France*, vols. 64; 66.
Vandel, A.	1965	Sur l'existence d'oniscoides très primitifs menant une vie aquatique et sur le polyphylétisme des isopodes terrestres. *Ann. Spéléologie* **20**, 489–518.

INDEX

Acknowledgements

Thanks are due to the following authors and publishers for kind permission to redraw figures from their publications:

from Edney 1953 (Linnean Society, London); figs. 25B, 33A(2).

from Gruner 1966 (VEB Gustav Fischer Verlag, Jena); figs. 22B, D; 25A, C, H; 31C; 33A(1), B, C, E, F, G.

from Harding 1969 (Entomologist's Monthly Magazine); fig. 33H.

from Schmölzer 1965 (Akademie-Verlag, Berlin); fig. 29E.

from Sheppard 1968 (Cave Research Group of G.B.); fig. 25E, F, G (F and G after Vandel, 1952).

from Vandel 1962 (Editions Paul Lechevalier, Paris); figs. 31D, E; 33D.

fig. 13 is after Gruner (unpublished) after Patané (1959).

fig. 15 is based on a figure in "The World of Spiders" by W. S. Bristowe 1958 (Collins, London), and on original observations.